Advanced Practical Chemistry

J S Clarke BSc MA CChem FRSC
Head of Science, Alleyn's School, London

S Clynes BSc CChem FRSC
Head of Science, The Manchester Grammar School

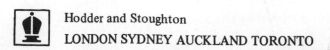
Hodder and Stoughton
LONDON SYDNEY AUCKLAND TORONTO

Clarke, John Shipley
   Advanced practical chemistry. — 2nd ed.
   1. Chemistry — Experiments
   I. Title II. Clynes — Solomon
   542   QD43

ISBN 0 340 24471 2

First printed 1973

Second edition 1979, Reprinted with revisions 1980, 1981.

Printed in Great Britain for
Hodder and Stoughton Educational,
a division of Hodder and Stoughton Ltd,
Mill Road, Dunton Green, Sevenoaks, Kent.
by J.W. Arrowsmith Ltd, Bristol

Summary of Contents

A detailed index to each Part is to be found on its first page.

Authors' preface

These experiments are for GCE 'A' level or equivalent courses. The order in each of the six Parts is similar to that found in books devoted to theoretical chemistry. There are more experiments here than can be performed in a two-year course, thus enabling selection to be made. In Part III traditional qualitative analysis by group separation has been given because it is useful background knowledge for many analytical exercises.

These experiments are for individuals (or pairs of students) to perform but it may be thought wiser to demonstrate some of them.

It is important that the theory relevant to all practical work is fully understood. Some of the theory for inorganic chemistry can be gained by studying Part II in conjunction with Part III and for organic chemistry by comparing Parts IV and V.

The IUPAC system of nomenclature and units has been followed. Some of the names suggested by ASE reports have been added.

Thanks for their assistance in the preparation and revision of these experiments are due to many colleagues, friends and students. We are always grateful to correspondents who offer further comments.

J.S.C.   S.C.

Part V: Suitable Unknowns for Analysis by Set Tests

5.27 A Propan–2–ol
5.28 B Butyraldehyde (butanal)
5.29 C Ethyl methyl ketone (butanone)
5.30 D Cinnamic acid (3-phenylpropenoic acid)
5.31 E Calcium formate (methanoate)
5.32 F Potassium oxalate (ethanedioate)
5.33 G Sodium acetate (ethanoate)
5.34 H Ammonium salicylate (2-hydroxybenzoate)
5.35 I Dimethyl oxalate (ethanedioate)
5.36 J Oxamide or urea oxalate (carbamide ethanedioate)
5.37 K Anilinium chloride (aniline hydrochloride) (phenylammonium chloride)

Part VI: Suitable Unknowns for Volumetric Analysis

6.5 ⎫ Sodium hydroxide 3 g/dm^3 with sodium
6.6 ⎭ carbonate-10-water 20 g/dm^3
6.10 Succinic (butanedioic) acid
6.12 0.2M hydrochloric acid containing 1.7 g/dm^3 magnesium carbonate and 2 g/dm^3 calcium carbonate
6.18 Dilute some 50-volume hydrogen peroxide one hundred times
6.20 Potassium trihydrogendioxalate-2-water
6.21 0.11M ammonium iron(II) sulphate-6-water; X = 0.05M hydrogen peroxide
6.28 3 cm^3 ethanol in 1 dm^3 aqueous solution coloured red by litmus and a drop of sulphuric acid
6.34 Hair bleach is about 0.2M hydrogen peroxide; dilute it five times; alternatively simulate it by diluting 50-volume hydrogen peroxide one hundred times
6.41 Sodium hydroxide 4 g/dm^3 with sodium chloride 6 g/dm^3
6.43 Mercury(I) nitrate-2-water 10 g/dm^3

In a laboratory

1 Follow the directions for your experiment very carefully; do not do anything else without obtaining permission.

2 Wear an overall to protect your normal clothing; objects such as rulers should not stick out of pockets; wash the overall frequently. If you do not wear spectacles, safety spectacles should be worn. Wash your hands thoroughly and often; rubber gloves may be advantageous for some operations.

3 Care should be taken with flammable materials when they are heated. If a fire starts turn off gas and electricity and also water if that would spread the fire; starve the fire of oxygen directly if possible or else use a dry powder or a carbon dioxide extinguisher.

4 The quickest and often the best treatment for a burn, scald or accidental contact with a corrosive chemical is washing with plenty of cold water. Report all accidents.

5 Be sure that the substances employed are those which are specified; solutions should be of the correct concentration. Use the quantities of materials advised. Keep your bench tidy; return stoppers to bottles and bottles to their storage places as soon as possible.

6 Many materials are poisonous: you must neither handle nor taste them. When smelling a substance proceed cautiously: waft the vapours towards your nose, sniff carefully, breathe in through your mouth and then expel the vapours from your nose by breathing down it. Use a fume cupboard whenever advised to do so; use a pipette filler always.

7 Concentrate on your own experiment and record all observations which you think are significant. Keep books etc. away from the working area of the bench if possible.

8 Use apparatus that is clean and in good condition. After the session clean and put away your apparatus tidily. Hand in apparatus you cannot clean and any surplus of the substances employed (do not put them directly into the stock bottle). Do not put solid materials that are insoluble in water nor organic solvents in a drain. Do not leave the gas, electricity or water supply on except by arrangement.

9 To cut a piece of glass tubing use a file to make a scratch, hold the tubing in a cloth or paper towel and bend the ends towards you, then anneal the new ends in a flame. To insert a thermometer or a piece of tubing through a cork, moisten it with water (for a rubber stopper use glycerol), hold the tubing in a cloth and gently insert it with a corkscrew motion.

10 Do not sit down to do experiments involving hot or corrosive materials. If you have to move around the laboratory do so at a reasonable pace, being careful not to disturb other people. Keep any bags or cases out of gangways.

---

11 A spatula load is taken to mean 0.25–0.5 g of material. A 100×16 mm test-tube can contain about $12 \, cm^3$; a 75×10 mm test-tube can contain about $4 \, cm^3$.

12 Concentrations of solutions are usually quoted in mol/dm^3, denoted by the symbol M. Molarity is a term which is not favoured in the SI scheme. The abbreviation m/V is used to indicate that the concentration of a solution is expressed in terms of a mass being present in a given volume.

Suggestions for the arrangement of written exercises

1 The account of a new experiment should always be started on a new page. The reference number (A.B) may be quoted in the margin and the date in the top right-hand corner. The title (central) should be clearly shown.

2 A line should be missed between each section of an exercise.

3 The margin should be left clear of everything except exercise numbers.

4 An experiment should be written up very carefully, stating the conditions, recording observations and deductions and arriving eventually at conclusions. Formulae should not be written as abbreviations in accounts.

5 A line should be missed before a sub-heading (under-lined) is used so that it may be seen clearly.

6 The end of an exercise should be shown by a ruled line on the next printed line.

Diagrams

Where appropriate an experiment should be illustrated by a clear diagram of a cross-section of the apparatus: this should be allowed the width of the page. It is advisable to use a pencil for diagrams so that corrections may be made.

Graphs

When a graph of A against B is required then A is the dependent variable and its values are put on the y axis and B is the independent variable and its values are put on the x axis. A graph shows the relationship of one set of numbers (quantities divided by units) to a second set of numbers. If the graph is a straight line the scales should be chosen so that the slope is approximately $45°$ or $135°$.

Suitable concentrations of solutions (mol/dm^3)

18   CH_3COOH (concentrated, glacial), NH_3 (concentrated, 880), H_2SO_4 (concentrated)

16   HNO_3 (concentrated), H_3PO_4 (concentrated, syrupy)

12   HCl (concentrated)

5   NaOH (concentrated)

2.5 $CaCl_2$ (for drying), KCl (saturated)

N.B. The above solutions are **not** used unless specifically requested.

1   CH_3COOH, $AlCl_3$, NH_3, $(NH_4)_2CO_3$, $(NH_4)_2SO_4$, $(NH_4)_2S$, $CaCl_2$, $CuCl_2$, $CuSO_4$, dimethylglyoxime (in ethanol), $C_6H_{12}O_6$ (glucose), HCl, H_2O_2, LiOH (including 0.5 M KNO_3), $MgCl_2$, $MgSO_4$, HNO_3, H_3PO_4, KOH, Na_2CO_3, NaOH, NaOCl, Na_2HPO_4, H_2SO_4, $ZnSO_4$

0.5 $Al_2(SO_4)_3$, KI, $SnCl_2$

0.1 $(NH_4)_2MoO_4$, $(NH_4)_2C_2O_4$, $BaCl_2$, Br_2 water, $Co(NO_3)_2$, $FeCl_3$, $FeSO_4$ [as $(NH_4)_2Fe(SO_4)_2$], $Pb(CH_3COO)_2$, $Pb(NO_3)_2$, $MnSO_4$, $HgCl_2$, $NiSO_4$, KBr, KCl, K_2CrO_4, $K_4Fe(CN)_6$, KI, KNCS, $AgNO_3$

0.025 $Ca(OH)_2$ (saturated), $CaSO_4$

0.02 $KMnO_4$

0.016 $K_2Cr_2O_7$

0.01   2,4–dinitrophenylhydrazine (in 3 M HCl), I_2 (in 0.02 M KI)

Others

Fehling's A	0.25 M $CuSO_4$
Fehling's B	2.5 M NaOH + 0.6 M potassium sodium tartrate
Hydrogen sulphide	0.5 cm^3 H_2O and 100 cm^3 CH_3COCH_3: blow in H_2S

I Physical Chemistry

The variation of water vapour pressure with temperature

Plot a graph for use in conjunction with your physical chemistry experiments of water vapour pressure (in mm Hg) against temperature (in the range 92–104°C). The values of water vapour pressure at low temperatures are also quoted but a graph is not needed because the variations are much smaller.

T	0	1	2	3	4	5	6	7	8
P	4.6	4.9	5.3	5.7	6.1	6.5	7.0	7.5	8.0

T	9	10	11	12	13	14	15	16	17
P	8.6	9.2	9.8	10.5	11.2	12.0	12.8	13.6	14.5

T	18	19	20	21	22	23	24	25
P	15.5	16.5	17.5	18.6	19.8	21.0	22.4	23.7

T	80	82	84	86	88	90	92	94
P	355	385	417	451	487	526	567	611

T	96	98	100	102	104	106
P	658	707	760	816	875	938

The normal distribution curve

This curve, discovered by De Moivre in 1733, has been given various names: Gaussian curve, the second law of Laplace, curve of the gendarme's hat, the normal probability distribution function, etc. The equation for the curve is

$$y = (2\pi)^{-\frac{1}{2}} \ e^{-x^2/2}$$

where y is the probability of the occurrence of a given value of x

and x is in units of the distance between the axis of symmetry and the point of steepest descent.

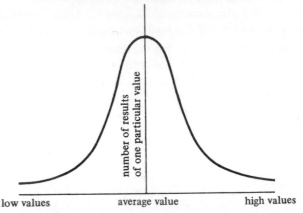

If one experiment is done several times by a student, or if one experiment is performed by several students the results obtained will always follow this pattern. This applies to all experiments and if the scatter of results does not follow this pattern then the validity of the experiment should be doubted.

The fraction of the total area of the graph lying within $x = \pm 1$ is 0.68 so that when a result is quoted as 25.06 ±0.02 it means that there is a 68% chance that a single experiment will give a value of the quantity between 25.08 and 25.04.

The Relative Molecular Mass of a Gas

1.1 Regnault's Method

1 Fit a flask of about 250–1000 cm³ capacity with a good stopper. Find the mass of the flask and the air contained in it (m_1 g).

2 Fill the flask with the dry gas, e.g. carbon dioxide made in a Kipp's apparatus from moderately concentrated hydrochloric acid and calcium carbonate, and dried by passing it through concentrated sulphuric acid. The delivery tube should reach to the bottom of the flask to ensure that all the air is displaced. Find the mass of the flask and the gas contained in it (m_2 g).

3 Find the total internal volume of the flask by filling it with water and finding the mass of the flask and water (m_3 g), if a balance of the necessary capacity is available. Alternatively, carefully pour the water out into measuring cylinders. The values of m_1 to m_3 are all less than the true values because of the upthrust of the air.

4 Record the temperature and the atmospheric pressure.

5 Water has a density of approximately 1 g/cm³ at a typical room temperature so the volume of the flask is $(m_3 - m_1)$ cm³ or as found from the cylinders. This volume must be corrected to standard temperature and pressure [273 K (0°C) and 760 mm Hg; s.t.p.]: V_0 cm³.

Air has a density of 0.0013 g/cm³ at s.t.p. so the mass of V_0 cm³ of air can be calculated (m_4 g). Hence the mass of V_0 cm³ of the gas is $m_2 - m_1 + m_4$ g.

6 A mole of a gas occupies 22.4 dm³ (22 400 cm³) at s.t.p. The answer should be accurate enough to decide what multiple of the relative empirical mass of the substance to take. What value of M_r do you obtain?

1.2 Dumas' Method

1 Find the mass of a Dumas' bulb and the air contained in it (m_1 g). It may be steadied if necessary on the balance by a piece of thread. Put about 10 cm³ of acetone (propanone) into the bulb by means of a teat pipette and put the bulb into a water bath up to its neck.

Rapidly heat up the bath and at the instant all the acetone has just vaporized seal off the bulb. The sealing-off should be done by heating the jet with a second burner and gripping the end of the jet with a pair of tongs: at the required instant pull with the tongs and the jet will 'neck' and seal. Keep the piece of glass which has become detached. Record the temperature of the waterbath: it should be about 30K (°C) above the boiling-point of the acetone (330K, 57°C).

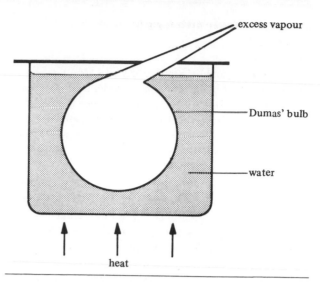

excess vapour
Dumas' bulb
water
heat

2 Allow the bulb to cool back to room temperature. It should be observed that the acetone within it will liquefy. Find the mass of the bulb containing the acetone together with the detached piece of glass (m_2 g).

3 With a glass cutter make a scratch on the glass jet about 1–2cm from the sealed end. Fill the sink with water and then break the sealed end off the jet by 'bending' the jet away from you. Some practice on scrap glass tubing is advised before tackling the jet. Water should enter and fill the bulb: this gives you a measure of how accurate your experiment is likely to be because it indicates how full of acetone vapour the bulb was at the instant of sealing. Find the mass of the bulb, water and two pieces of glass, if a balance of the necessary capacity is available (m_3 g). Alternatively, carefully pour the water out into measuring cylinders. The values of m_1 to m_3 are all less than the true values because of the upthrust of the air.

4 Record the temperature of the laboratory and the atmospheric pressure.

5 Water has a density of approximately 1 g/cm³ so the volume of the flask is $(m_3 - m_1)$ cm³. This volume must be corrected to s.t.p. (V_1 cm³).

Air has a density of 0.0013 g/cm³ at s.t.p. so the mass of V_1 cm³ of air can be calculated (m_4 g).

Neglecting the thermal expansion of the bulb, $(m_3 - m_1)$ cm³ of vapour were sealed in the bulb at the bath temperature and atmospheric pressure: so this volume can also be corrected to s.t.p. (V_2 cm³). The mass of gas sealed in the bulb was $m_2 - m_1 + m_4$ g.

6 A mole of a gas occupies 22.4 dm³ at s.t.p. The answer should be accurate enough to decide what multiple of the relative empirical mass of the substance to take. What value of M_r do you obtain?

1.3 Victor Meyer's Method

1 Warm a little sand on a metal tray and then put the dried sand to a depth of 5 mm into the inner tube to cushion the fall of the Hofmann weighing bottle. Heat the inner tube in a vapour (steam) bath for about 20 minutes until the temperature is steady, as shown by the lack of any bubbles in the beaker of water into which the side-arm dips.

2 Find the mass of a Hofmann bottle empty and then full of the liquid under test, e.g. diethyl ether (ethoxyethane). Handle the bottle as little as possible, and then only by the rim, to minimize losses by evaporation. Remove the stopper from the inner tube for as brief a time as possible in order to drop the Hofmann bottle in. Immediately move the eudiometer tube (or an inverted burette) over the delivery tube. A subsidiary experiment enables the volume between the bottom graduation and tap on the burette to be found: run in water from a second burette. The stopper should blow out of the Hofmann bottle and ether vapour will displace air out of the inner tube into the eudiometer.

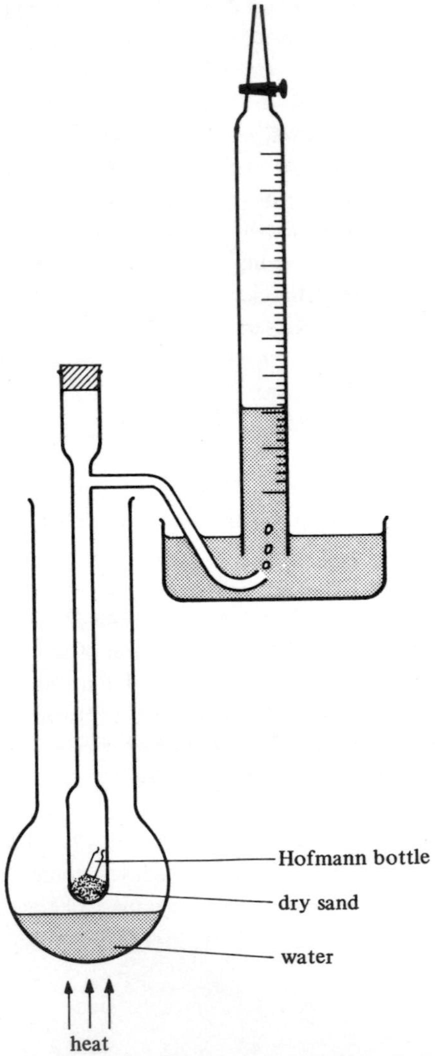

3 When no more air comes out into the eudiometer move it carefully to a large measuring cylinder (or bucket) and allow the air to cool to room temperature. Adjust the height of the eudiometer to make the levels of water inside and outside the same: hence the pressure of the air and water vapour in the eudiometer is equal to the atmospheric pressure.

4 Record the temperature of the water in the measuring cylinder and the atmospheric pressure. Look up the water vapour pressure at this temperature.

5 The volume of air (and hence vapour of the substance) must be corrected to s.t.p.

6 A mole of a gas occupies 22.4 dm^3 at s.t.p. The answer should be accurate enough to decide what multiple of the relative empirical mass of the substance to take. What value of M_r do you obtain?

Hofmann bottle

dry sand

water

heat

Introduction

Bridges for measuring the conductance of solutions are available, e.g. Grayshaw CT 50 Conductance Bridge, so that the experimental set-up of a Wheatstone Bridge is not required.

When the bridge is switched on, the pointer of the small meter should swing to the right. The value of the conductance of the solution in siemens (S, ohm^{-1}) is obtained by multiplying the reading on the main dial (1 to 12) by that shown on the range switch (1 to 10^{-7}). A resistor should be used for practice: the range switch is turned to the expected setting and the main dial slowly rotated until the meter pointer drops towards the left or zero position. The point of balance of the bridge is achieved when the pointer reaches a minimum; it does not always reach zero. Then the resistor must be removed.

The conductivity cell is usually of the dipping variety; it is very fragile and expensive. It consists of two pieces of platinum-black fixed to the opposite sides of a glass cylinder. Sufficient of the solution under investigation must be put into a small beaker to cover the electrodes. Between readings, the cell should be rinsed with distilled water and allowed to drain dry.

1.4 The Comparison of the Conductance of Solutions and Liquids

Using the conductance bridge, measure the conductance of the following.

1. hydrochloric acid, acetic (ethanoic) acid, sodium hydroxide and ammonia in dilute solution.

2. tap-water and distilled water.

3. ethanol, glacial acetic (ethanoic) acid, concentrated sulphuric acid, toluene (methylbenzene) and nitrobenzene.

Comment upon the results you obtain.

1.5 The Variation of Molar Conductivity with Dilution

The concentrated hydrochloric acid supplied to the laboratory is about 11–12 M: probably it will have been standardized in a volumetric exercise (see experiments 6.1, 6.2, 6.40 and 6.45). The dilutions must be carried out carefully using a burette or pipette and a measuring flask and at each stage the solution should be shaken thoroughly to ensure that it is homogeneous. Typical concentrations at which to measure the conductance are approximately 12, 6, 3, 1.5, 0.75, 0.38, 0.19, 0.095 and 0.048 M.

An alternative substance is glacial acetic (ethanoic) acid.

The conductance (G) of a solution when multiplied by the cell constant (distance apart of electrodes divided by area of cross-section of an electrode) gives the electrolytic conductivity (κ, in S m^{-1}) of the solution. The electrolytic conductivity must then be divided by the concentration of the solution in mol/m^3 of electrolyte to give the molar conductivity (Λ in S m^2 mol^{-1}).

Two graphs should be plotted, one of molar conductivity against concentration and one of molar conductivity against the square root of the concentration. Comment on the shapes of the graphs obtained.

1.6 A Conductimetric Titration

To minimize changes in conductivity with dilution a titration is performed with a dilute solution and some water in the conical flask and a concentrated solution in the burette. Compare this with volumetric analysis where the concentrations are adjusted so that the volumes of solutions titrated are roughly equal.

Put 25 cm^3 of 0.1 M sodium hydroxide solution and about 100 cm^3 of distilled water in a 250 cm^3 conical flask, or a beaker, over a magnetic stirrer (if one is available). In a burette place 0.5 M hydrochloric acid. Measure the conductance of the alkali at the start and after the successive additions of ten 1 cm^3 portions of the acid.

Plot a graph of conductance against volume of acid added. Comment upon its shape and check whether the solutions have the relative concentrations stated. The titration should also be performed using phenolphthalein or screened methyl orange as the indicator.

An alternative pair of substances is saturated barium hydroxide solution (25 cm^3) and 0.5 M sulphuric acid.

1.7 The Determination of a Solubility Product

Add about 10 cm^3 of 0.1 M hydrochloric acid to 10 cm^3 of 0.1 M silver nitrate solution and centrifuge (or filter) off the precipitate. Wash the precipitate thoroughly with distilled water several times. Then shake the precipitate with about 50 cm^3 of distilled water in a clean vessel. Centrifuge the solution and measure the conductance. Calculate the electrolytic conductivity of the solution.

The molar conductivity of Ag$^+$ is 6.2×10^{-3} S m^2 mol^{-1} and of Cl$^-$ is 7.6×10^{-3} S m^2 mol^{-1}.

Let the solubility of silver chloride be x mol/dm^3. Calculate the value of x and hence the solubility product. What are the units of the solubility product in this example?

1.8 The Hydrogen and Calomel Electrodes

A hydrogen electrode is made by bubbling hydrogen at one atmosphere (101 kPa) pressure over platinum-black on the surface of which hydrogen atoms form. The platinum is immersed in 1 M hydrochloric acid and the rate of passing bubbles is about one each second.

The platinum electrode is prepared by first using it as the cathode for electrolyzing a 3% m/V solution of tetrachloroplatinic(II) acid containing $5 cm^3$ of 0.05 M lead acetate (ethanoate) in every 1 dm^3 of solution. The current density should be about 15 mA/cm^2 and the electrolysis continued until the silver colour of the platinum vanishes: after about five minutes a matt black surface is obtained. Secondly the electrode should be used as the cathode in the electrolysis of dilute sulphuric acid for two minutes. The platinum-black electrode must be washed thoroughly with distilled water and kept in water when not in use. The hydrogen electrode is the standard and other reduction potentials are measured relative to it.

The calomel electrode is a useful secondary standard. In the calomel electrode a platinum wire makes the electrical connection from the external circuit to some mercury. On top of the mercury rests some mercury(I) chloride (calomel, Hg_2Cl_2), and the solid is in equilibrium with a saturated solution of mercury(I) chloride in a solution of potassium chloride.

To make a connection from one electrolyte (half cell) to another, a salt-bridge is used; this is an n-shaped tube full of concentrated potassium chloride or nitrate solution, having loose plugs of glass-wool at each end to prevent solutions mixing too much.

Set up hydrogen and calomel electrodes connected by a salt-bridge. Using a high resistance voltmeter (e.g. a valve voltmeter) connected to the two electrodes, measure the electromotive force of the calomel electrode. The answer depends on the concentration of the potassium chloride solution, e.g. at 298 K (25°C):

0.1 M	−0.335 V
1 M	−0.280 V
saturated	−0.242 V

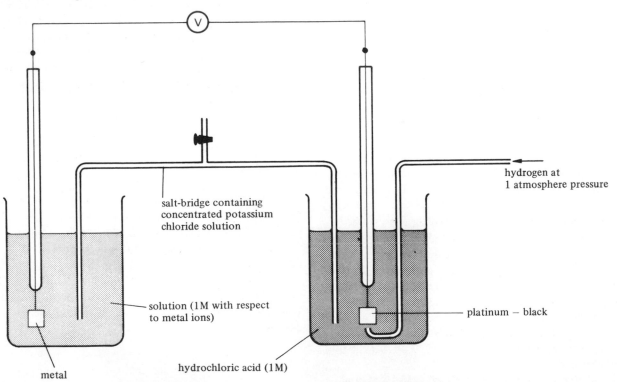

salt-bridge containing concentrated potassium chloride solution

hydrogen at 1 atmosphere pressure

solution (1M with respect to metal ions)

platinum − black

metal

hydrochloric acid (1M)

1.9 The Determination of Relative Electrode Potentials

Using a hydrogen or calomel electrode as your reference measure the relative electrode potentials of a series of metals. In each case a 1M solution of the metal salt should be in contact with a clean strip of the metal, e.g.
a) copper and copper(II) sulphate solution,
b) zinc and zinc sulphate solution,
c) magnesium and magnesium sulphate solution,
d) nickel and nickel sulphate solution.

In some cases the metal is more electropositive (forms positive ions more readily) than hydrogen, in other cases it is less so. Care is needed because the voltmeter may be connected the wrong way round. By convention the electromotive forces measured refer to the cells:

$$Pt, H_2 \mid H^+ \mid ion\ M^+ \mid metal\ M$$

where the electrode on the left is the standard hydrogen electrode and the e.m.f. measured is the relative electrode potential. This implies the reaction:

$$\tfrac{1}{2}H_2 + M^+ \rightarrow H^+ + M$$

It is thus a reduction potential:

$$M^+ + e^- \rightarrow M$$

and is negative for a metal that forms ions more readily than hydrogen in aqueous solution.

1.10 The Effect of Concentration on Relative Electrode Potential

Set up the cell:

$$Pt, H_2 \mid H^+ \mid Ag^+ \mid Ag$$

using a concentrated solution of silver nitrate. The connection between the solutions should be made by means of a salt-bridge containing potassium nitrate. Measure the relative electrode potential (E) of the solution. Using a burette or pipette and a measuring flask dilute the solution accurately, and again measure the potential; repeat this several times.

Plot a graph of E against $\log_{10} [Ag^+]$; measure the slope of the graph and the intercept on the y axis.

The values obtained should be compared with those obtained using the equation:

$$E = E^\ominus + \frac{RT}{zF} \log_e [Ag^+]$$

The standard electrode potential ($E^\ominus$) of silver is + 0.80V; R is the gas constant, T the thermodynamic temperature, z the change in valency of the ion and F the Faraday constant.

$$\log_e [Ag^+] = 2.303 \log_{10} [Ag^+]$$

1.11 A Potentiometric Titration

Set up the cell:

$$Pt, H_2 \mid H^+ \mid Fe^{2+}, Fe^{3+} \mid Fe$$

using 10cm^3 portions of 0.2M solutions of ammonium iron(II) sulphate and ammonium iron(III) sulphate. The connection between the acid and iron solutions should be made by means of a salt-bridge.

Fill a burette with 0.02M potassium permanganate [manganate(VII)] solution. Measure the relative electrode potential of the iron solution. Add 5cm^3 portions of potassium permanganate solution up to a maximum of 40cm^3 to the iron solution, measuring the potential after each addition.

Plot a graph of the potential against the volume of permanganate added. Discuss the shape of the graph. At what stage did the iron solution become pink?

1.12 A Redox Problem

You are provided with solutions (about 0.1 M) of three oxidizing agents:
a) potassium permanganate [manganate(VII)], which is acidified with dilute sulphuric acid,
b) sodium hypochlorite [chlorate(I)],
c) potassium chlorate [chlorate (V)],
and three reducing agents:
d) ammonium iron(II) sulphate [sulphate(VI)],
e) sodium sulphite [sulphate(IV)],
f) hydrazinium hydrogensulphate [sulphate(VI)] $N_2H_5^+ HSO_4^-$,
and also with some hydrogen peroxide (use a 2-volume solution).

Decide what are the reduction products of the oxidizing agents and the oxidation products of the reducing agents. Establish tests for the presence or absence of all these substances.

Mix each of the oxidizing agents with each of the reducing agents (3 cm^3 portions are suitable) and check whether a redox reaction has occurred. Put hydrogen peroxide with each of the substances (a) to (f) and check whether a redox reaction has occurred. From your observations place the seven substances in an order such that the best oxidizing agent is at the top and the best reducing agent at the bottom.

Look up the reduction potentials of each of the substances and comment on whether a reaction was to be expected.

$$[N_2H_5^+ + e^- \rightarrow N_2 + \tfrac{5}{2}H_2; E^\ominus = -0.5\ V]$$

1.13 Enthalpy of Mixing

The forces between molecules of different substances may vary widely; they may be attractive or repulsive, or approximately the same as between the molecules of the separate species. Some idea of their nature may be obtained by measuring the enthalpy (heat) of mixing of the two species.

1   Put 22 cm^3 of paraxylene (1,4-dimethylbenzene, density 0.87 g/cm^3) in a beaker and record its temperature. Into a second beaker put 25 cm^3 of toluene (methylbenzene, density 0.87 g/cm^3) and record its temperature. Pour the xylene rapidly into the toluene, stir well and observe the alteration (if any) in the temperature.
    Taking the speciifc heat capacity of the mixture as 1.6 J g^{-1} K^{-1}, calculate the enthalpy of mixing one mole of xylene with one mole of toluene. The hydrocarbon mixture should be put in the appropriate residues bottle. (**Care**: use a fume cupboard.)

2   Repeat experiment (1) using 58 cm^3 of ethanol (density 0.79 g/cm^3) and 18 cm^3 water (density 1 g/cm^3); take the specific heat capacity of the mixture as 3.33 J g^{-1} K^{-1}.
    Calculate the enthalpy of mixing one mole of ethanol with one mole of water.

3   Put 20 cm^3 of concentrated hydrochloric acid into the conical flask and run in 30 cm^3 of concentrated sulphuric acid from the tap-funnel. (**Care**: concentrated acids). Collect the hydrogen chloride evolved in 54 cm^3 of water by the inverted funnel technique as shown in the figure.

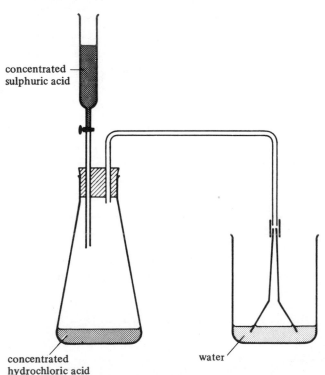

concentrated
sulphuric acid

concentrated
hydrochloric acid

water

Measure the variation in temperature of the contents of the beaker. Using a balance, determine the increase in mass of the water.
    Taking the specific heat capacity of the water as 4.2 J g^{-1} K^{-1}, calculate the enthalpy of mixing one mole of hydrogen chloride with three moles of water. (This does not include any allowance for the enthalpy dilution of a solution.)

Comment upon the values obtained in experiments (1) to (3).

1.14 Enthalpy of Neutralization

1   Put 50 cm^3 3 M sodium hydroxide solution in a beaker and record its temperature. Put 50 cm^3 of 1 M hydrochloric acid in a second beaker and record its temperature.
    Pour the acid into the alkali, stir well and record the maximum temperature attained.
    Taking the specific heat capacity of the mixture as 4.2 J g^{-1} K^{-1}, calculate the enthalpy of neutralization for one mole of hydrochloric acid.
    The experiment may be performed using bottles embedded in polystyrene blocks to insulate them, but the heat losses and the heat capacity of the apparatus are negligible in this rapid experiment.

2   Repeat experiment (1) using 1 M sulphuric, nitric, acetic (ethanoic) and phosphoric acid.

3   The experiment may also be repeated using 1 M solution of ammonia with hydrochloric acid, and ammonia with acetic (ethanoic) acid.

Comment upon the values obtained in your experiments.

1.15 Enthalpy of Combustion

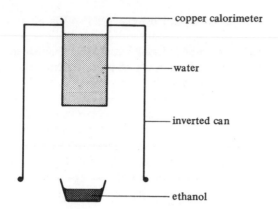

copper calorimeter

water

inverted can

ethanol

The top of the crucible should be level with the bottom
of the inverted can. Find the mass of the calorimeter
and can. Put 100 cm³ of water in the copper calorimeter
and measure its temperature. Shield the can from draughts
as much as possible with hardboard sheets.

In a small crucible burn a known mass of ethanol
(2 cm³, density 0.79 g/cm³). Measure the maximum
temperature attained by the calorimeter and the water.

The energy evolved on burning 2 cm³ of ethanol is
given by the product of: the mass of water, the specific
heat capacity of water (4.2 J g⁻¹ K⁻¹) and the increase in
temperature. To this add the product of: the mass of
the calorimeter and can, the specific heat capacity
(0.42 J g⁻¹ K⁻¹) and the increase in temperature.
Hence calculate the enthalpy of combustion of one
mole of ethanol.

The experiment gives a lower answer than is recorded
in text-books because of heat losses to the surroundings
but it may be used to compare the enthalpies of combustion of liquids which burn readily.

1.16 The Distillation of Completely Miscible Liquids

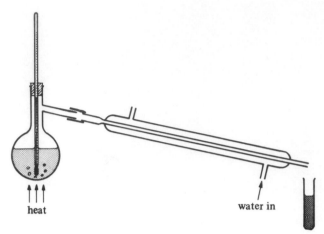

heat   water in

1 Set up the apparatus as shown in the figure. The thermometer ($0–110\,^{\circ}$C) must project almost to the bottom of the flask ($250\,cm^3$). Mark 15 clean test-tubes so that $10\,cm^3$ of liquid can be collected in each. Put exactly $150\,cm^3$ of 1M hydrochloric acid, i.e. 150 millimoles, into the flask and sprinkle in a few anti-bumping granules to ensure smooth ebullition.

2 Heat strongly at first to bring the acid to the boil, and then moderately but constantly until the flask has almost run dry. Read the thermometer at the beginning and the end of the collection of each $10\,cm^3$ sample of distillate. Calculate the mean temperature of distillation for each sample.

3 Pour each sample into a conical flask, rinse the test-tube twice with water and add the rinsings to the sample: titrate each sample with 1M sodium hydroxide solution using screened methyl orange as the indicator. The titres are *very* low until towards the end of the experiment.

4 The results should be recorded on a table:
    serial number of distillate,
    temperature at start,
    temperature at end,
    average temperature
    titre of distillate,
    concentration of acid in distillate (M_1),
    total number of millimoles of acid distilled,
    number of millimoles of acid left in flask,
    volume of acid left in flask,
    concentration of acid in flask (M_2).

On the same graph plot the average temperature (vertical axis) against the concentration of the acid in the flask and in the distillate (both on horizontal axis). From the intersection of the two curves obtained deduce the concentration and temperature of the constant boiling-point mixture.

1.17 A Study of Two Partially Miscible Substances

Care: phenol is **very** corrosive.

1 Fit a boiling tube ($150 \times 25\,mm$) with a rubber stopper and weigh into it about 7g of phenol accurately. Add $3\,cm^3$ of distilled water from a burette.

2 Remove the stopper for this stage; use a $0–110\,^{\circ}$C thermometer and a stirrer. Clamp the boiling-tube so that the substances in it are immersed in a beaker of water which may be heated. Heat the beaker of water until the mixture in the boiling tube becomes clear, i.e. the substances have mixed, and record the temperature.

    Allow the boiling tube to cool by removing it from the beaker of water, keep the phenol and water well stirred, and record the temperature at which the mixture becomes cloudy again. The average of the two temperatures is the temperature at which 7g of phenol is soluble in 3g of water, or vice versa.

3 Run in $1\,cm^3$ of water from the burette and repeat the heating and cooling as in (2). Continue adding water in $1\,cm^3$ portions until the consequent average temperature has reached a maximum (and perhaps subsequently decreased).

4 Take $9\,cm^3$ of water in a second boiling tube and weigh in about 1g of phenol accurately. Repeat the heating and cooling as in (2). Add further 1g portions of phenol and find the corresponding temperatures as in step (3).

5 Carefully record all masses and temperatures as the measurements are made. Draw up a table from these measurements of:
    mass of phenol,
    mass of water,
    percentage of phenol,
    percentage of water,
    mean temperature.

Plot a graph of the mean temperature (vertical axis) against the percentage of phenol (horizontal axis). Lightly shade in the area on the graph corresponding to the observation of a cloudy liquid (heterogeneous region).

[If a pair of students do the experiment then one should start at (1) and the other at (4).]

1.18 The Steam Distillation of Immiscible Liquids

1 Set up the apparatus shown in the figure. Put hot
   water into the steam generator to expedite progress.
   Into the bolt-head flask pour by measuring cylinder
   about 100 cm³ of water and about 40 cm³ of chlorobenzene (density 1.1 g/cm³). The thermometer must
   be immersed in the liquids.

2 Heat the steam generator and the bolt-head flask, but
   when the temperature of the latter reaches about
   350 K (77°C) turn off its Bunsen burner.

3 Record the temperature as distillation proceeds; this
   should be fairly constant at about 365 K (92°C).
   Maintain the passage of steam and collect the distillate
   until about 30 cm³ of the chlorobenzene has been distilled. Disconnect the steam generator to obviate
   'sucking-back' upon cooling.

4 Record the volumes of chlorobenzene and water
   collected, the atmospheric pressure and the water
   vapour pressure at the temperature of distillation (see
   the graph plotted at the start of Part I).

5 Using the Naumann equation, which can be derived
   from the general gas equation, calculate the relative
   molecular mass of chlorobenzene:

$$\frac{\text{Vapour pressure of chlorobenzene}}{\text{Vapour pressure of water}} =$$

$$\frac{\text{Number of moles of chlorobenzene distilled}}{\text{Number of moles of water distilled}}$$

   The answer should be accurate enough to decide what
   multiple of the relative empirical mass to take so as to
   give the relative molecular mass. Is it?
      The wet chlorobenzene can be returned for future
   use.

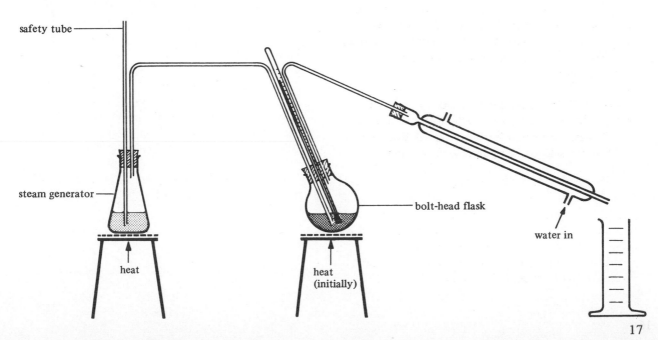

safety tube

steam generator

bolt-head flask

water in

heat

heat
(initially)

1.19 To Determine a Distribution Constant

Care: diethyl ether (ethoxyethane) is very flammable and care should be taken to eliminate flames in the vicinity. A pipette filler must be used.

1   Check that the tap on a separating funnel ($100\,cm^3$) is well-greased and that the stopper is efficient. Put about $30\,cm^3$ of water (by measuring cylinder) into the funnel and then add about $30\,cm^3$ of diethyl ether (the same measuring cylinder may be used). It is advantageous to put the separating funnel in a stand over a large beaker in case of accidents. Weigh in about $0.5\,g$ of powdered succinic acid (butanedioic acid) – the mass does not need to be known accurately. Shake the funnel gently, invert it and, by opening the tap, release the pressure that has built up. The enthalpy of dissolution of the acid causes some vaporization of the solvents. The shaking must be continued until all the crystals have dissolved and there is no hiss upon opening the tap.

2   Remove the stopper of the funnel and run off the lower (water) layer into a clean dry beaker. The boundary layer can be discarded and then the upper (ethereal) layer run into another beaker. For the titration a supply of approximately $0.5\,M$ sodium hydroxide solution is required; some of it should be diluted accurately five times using a pipette and measuring flask ($20\,cm^3$ to $100\,cm^3$ solution).

   Take $10\,cm^3$ of each layer in separate conical flasks; to the ethereal portion add approximately $10\,cm^3$ of water to facilitate titration. There should be enough of each layer to do each titration twice. Titrate the water layer directly with $0.5\,M$ sodium hydroxide using phenolphthalein as the indicator ($T_1\,cm^3$): proceed until the solution is a permanent pale pink, which may fade after five minutes because of the effect of atmospheric carbon dioxide. Titrate the ethereal layer with the $0.1\,M$ sodium hydroxide solution ($T_2\,cm^3$): swirl but do not shake the flask as you proceed because the neutralization only occurs in the aqueous part.

3   Repeat (1) and (2) using next about $1\,g$ of crystals and finally about $1.5\,g$. Record the temperature at which these experiments have been performed. Rinse out the burette with dilute hydrochloric acid and then water.

4   The volumes of $0.1\,M$ sodium hydroxide solution, which could have been employed for both titrations, i.e. $5\,T_1$ and $T_2\,cm^3$ respectively, are directly proportional to the concentrations of acid in each layer.

   If the distribution law is true then a graph of T_2 (vertical axis) against T_1 (horizontal axis) is a straight line of slope $5K$ where K is the distribution constant. The origin is a point on the graph. From your graph calculate the distribution constant of succinic acid between diethyl ether and water at room temperature.

1.20 The Dimerization of Carboxylic Acids

Care: glacial acetic (ethanoic) acid is very corrosive, and carbon tetrachloride (tetrachloromethane) has an obnoxious vapour: a pipette filler must be used.

---

1 Check that the tap on the separating funnel ($100 cm^3$) is well greased and that the stopper is efficient. Put about $30 cm^3$ of tetrachloromethane (by measuring cylinder) into the funnel and then add about $30 cm^3$ of distilled water (the same measuring cylinder may be used). It is advantageous to put the separating funnel in a stand over a large beaker in case of accidents. From a communal burette add about $1.5 cm^3$ of glacial acetic acid. Shake the funnel gently, invert it and by opening the tap release the pressure that has built up. The enthalpy of dissolution of the acid causes some vaporization of the solvents. The shaking must be continued until there is no hiss upon opening the tap.

---

2 Remove the stopper of the funnel and run off the lower (tetrachloromethane) layer into a clean dry beaker. The boundary layer can be discarded and then the upper (water) layer run into another beaker. For the titration a supply of 0.5 M sodium hydroxide solution is required; some of it should be diluted accurately five times using a pipette and measuring flask ($20 cm^3$ to $100 cm^3$ solution).

Take $10 cm^3$ of each layer in separate conical flasks; to the tetrachloromethane portion add approximately $10 cm^3$ of water to facilitate titration. There should be enough of each layer to do each titration twice.

Titrate the water layer directly with 0.5 M sodium hydroxide solution using phenolphthalein as the indicator ($T_1 cm^3$): proceed until the solution is a permanent pale pink. This may fade after five minutes because of the effect of atmospheric carbon dioxide.

Titrate the tetrachloromethane layer with the 0.1 M sodium hydroxide solution ($T_2 cm^3$): swirl, not shake, the flask as you proceed because the neutralization only occurs in the aqueous part.

---

3 Repeat steps (1) and (2) using next about $3 cm^3$ of glacial acetic acid and finally about $4.5 cm^3$. Record the temperature at which these experiments have been performed. Rinse out the burette with dilute hydrochloric acid and then water.

4 The volumes of 0.1 M sodium hydroxide solution, which could have been employed for both titrations, i.e. $5T_1$ and T_2 respectively, are only directly proportional to the concentrations of acid in each layer if the acid exists in the same molecular state in each layer.

Plot a graph of T_2 (vertical axis) against T_1 (horizontal axis) the origin being a point on the graph: it is not a straight line because the acid has dimerized in the tetrachloromethane.

From the titres calculate the concentration (in mol/dm^3) of the acetic acid in each layer, remembering that there are double molecules in the tetrachloromethane layer. Plot a graph of the square root of the concentration of the acid in the tetrachloromethane layer (vertical axis) against the concentration in the aqueous layer (horizontal axis). Again the origin is a point on the graph; the straight line produced proves that the acid has dimerized in the non-aqueous layer.

A straight line could also be obtained by plotting the concentration in the tetrachloromethane layer against the square of the concentration in the aqueous layer but this would correspond to complete ionization of the acetic acid in the water, a suggestion that is not supported by other experiments.

Alternative experiment: 1,1,1-trichloroethane can be used instead of carbon tetrachloride; the sodium hydroxide solution for titrating this layer should be 0.05 M.

1.21 To Determine the Formula of the Complex Ion from Copper(II) Ion and Ammonia Molecules

Care: a pipette filler must be used for chloroform (trichloromethane).

1 Check that the tap on a separating funnel ($100 \, cm^3$) is well-greased and that the stopper is efficient. Put about $30 \, cm^3$ of 1M ammonia solution (by measuring cylinder) into the funnel and then add about $30 \, cm^3$ of trichloromethane (the same measuring cylinder may be used). It is advantageous to put the separating funnel in a stand over a large beaker in case of accidents. Shake the funnel gently, invert it and by opening the tap release the pressure that has built up. The enthalpy of dissolution of the ammonia in the trichloromethane may cause some vaporization. The shaking must be continued until there is no hiss upon opening the tap.

2 Remove the stopper of the funnel and run off the lower (trichloromethane) layer into a clean dry beaker. The boundary layer can be discarded and then the upper (water) layer run into another beaker. For the titration a supply of 0.5M hydrochloric acid is required; some of it should be diluted accurately ten times using a pipette and measuring flask ($10 \, cm^3$ to $100 \, cm^3$ solution).

 Take $10 \, cm^3$ of each layer in separate conical flasks; to the trichloromethane portion add approximately $10 \, cm^3$ of water to facilitate titration. There should be enough of each layer to do each titration twice. Titrate the water layer directly with 0.5M hydrochloric acid, using methyl orange as the indicator.

 Titrate the trichloromethane layer with the 0.05M hydrochloric acid: swirl the flask as you proceed because the neutralization only occurs in the aqueous part. Record the temperature. The distribution constant of ammonia between trichloromethane and water should be calculated from the titres.

3 By pipettes put $20 \, cm^3$ 1M ammonia solution, $20 \, cm^3$ 0.1M copper(II) sulphate solution and $40 \, cm^3$ trichloromethane in the separating funnel. After shaking and separating, only the trichloromethane layer can be titrated: use $25 \, cm^3$ of this layer and repeat this section of the experiment once or twice.

4 From the average titration in (3) calculate the number of millimoles of ammonia in the trichloromethane layer. Using the distribution constant found in (2) calculate the number of millimoles of free ammonia in the aqueous layer. $20 \, cm^3$ 1M ammonia solution were used in (3) so the total number of millimoles of ammonia present is 20. By subtraction the number of millimoles of ammonia in combination

with the copper ions is found. The number of millimoles of copper ions is 20 x 0.1, i.e. 2. Calculate the ratio of the number of millimoles of ammonia to one of copper and hence obtain the formula of the complex ion.

 By a similar experiment using potassium iodide solution, iodine and tetrachloromethane, the formula of the complex ion formed by iodine can be found.

1.22 To Determine the Composition of a Eutectic Mixture

1 A test-tube can be held concentric in a boiling-tube by means of a cork ring at its upper end: the gap between the two tubes is an air-jacket which serves to reduce the rate of change in temperature. A beaker of water can be used to heat the boiling-tube. At the melting-point of a solid the crystals become pasty and opaque.

2 Accurately weigh out about 1 g of 2-nitrophenol into the test-tube and determine the melting-point of the crystals. Add 0.2 g portions of 4-nitrotoluene (methyl-4-nitrobenzene) by off-loading it from a watch-glass holding about 0.6g. The mass of each portion should be found by subsequent weighing of the watch-glass and crystals. Find the temperature at which crystallization starts for each molten mixture.

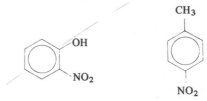

2-nitrophenol   *4-nitrotoluene*

3. Accurately weigh out about 1 g of 4-nitrotoluene into another test-tube set up as before and determine the melting-point of the crystals. Add 0.2 g portions of 2-nitrophenol by off-loading it from a watch-glass holding about 0.6g. Proceed as in (2).

4 Plot a graph of melting-point against percentage of 2-nitrophenol in the mixture. Deduce the eutectic temperature and the composition of the eutectic mixture. Does the percentage by mass correspond to any simple molecular ratio? A simple eutectic mixture will give a V-shaped graph; if there is definite compound formation a W-shaped graph is obtained.

5 A third test-tube can be used to contain mixtures of about the eutectic composition in order to clarify and verify the central section of the graph.

[If the experiment is done by a pair of students then one starts with (1) and (2) and the other with (1) and (3)].

1.23 The Properties of Ion Exchange Resins

1 The capacity of a resin

This can be found for an anion exchange resin, which may be a quaternary ammonium strong base in the form of its chloride. A neutral solution of a nitrate is used to displace the chloride ions from the resin and then a titration with standard silver nitrate solution performed.

Weigh out accurately about 1 g of the resin and put it in a conical flask with about 25 cm³ of distilled water. Add about 4 g of pure potassium nitrate and stir the mixture well. Then add 0.5 cm³ of potassium chromate solution as an indicator and titrate the solution with standard, e.g. 0.1 M silver nitrate solution. The mixture must be stirred well because all the chloride ions from the resin may not have been displaced before the titration starts and this leads to the apparent end-point being reversed several times before becoming permanent. Calculate the capacity of the resin in moles of chloride ion for 1 kg of resin.

2 The rate of exchange

The time factor in ion exchange mentioned above can be investigated by using a cation exchange resin, which may be a sulphonated polystyrene: usually supplied in the form of its sodium salt.

Accurately weigh out about 2 g of the resin and put it in a filter-paper in a funnel. The resin is converted to its hydrogen form by washing it twice with two bed-volumes of moderately concentrated (one volume concentrated acid: three volumes of water, i.e. about 3 M) hydrochloric acid and then with distilled water until the filtrate emerging from the paper is neutral. The resin should then be washed into a 250 cm³ beaker, stirred well (placed on a magnetic stirrer if one is available) and the volume of water made up to about 50 cm³. Add a few drops of methyl orange to serve as an indicator and fill a burette with standard, e.g. 0.1 M sodium hydroxide solution. At a convenient time, designated zero, add 10 cm³ of 0.1 M sodium chloride solution to the resin and water.

The reaction that occurs is

$$R^-H^+ \quad + \quad Na^+ \rightarrow R^-Na^+ + H^+$$

(R = resin)   (from NaCl)

After one minute run in alkali from the burette until the indicator changes colour (the titre is fairly low). Let the exchange continue and continue the titration at times such as 2, 3, 4, 5 . . . minutes until no further alkali is required.

Plot a graph of the total titre against time. Compare the shape of the graph with that obtained by the hydrolysis of an ester (experiment 1.43), or of a sugar (experiment 1.44).

3 The dissolution of a precipitate

Put about 15 cm³ of distilled water into a boiling tube (150 x 25 mm); add two drops of lead acetate (ethanoate) solution (0.25 M) and two drops of potassium iodide solution (0.5 M). Observe what happens and then add about 10 cm³ of a cation exchange resin in its sodium form. Shake the mixture well, observe what happens and explain it.

1.24 The Demineralization and Softening of Water

These two experiments can be done concurrently: proceed with part (2) while solutions are running through the columns.

1. The two stage demineralization of water

Put about 20 cm³ of a cation exchange resin, in its sodium form, into a column; convert it into its hydrogen form by washing it with 30 cm³ of moderately concentrated hydrochloric acid (3 M) and then rinse the column with about 75 cm³ of distilled water. Both these operations should be carried out at a rate of about 5–10 cm³/min. The washings should then be neutral. *Never permit a column to run dry.*

Put about 20 cm³ of an anion exchange resin, in its chloride form, into a column; convert it into its hydroxide form by washing it with 50 cm³ of 1 M sodium hydroxide solution and then rinse the column with about 150 cm³ of distilled water, which has been freed from carbon dioxide by warming and allowing it to cool out of contact with air. The washings should then be neutral.

Dissolve about 1 g each of copper(II) sulphate-5-water and potassium dichromate [dichromate(VI)] crystals in about 250 cm³ of water and run the solution through the cation and anion exchange columns at the same rate as above. Explain the colour changes observed on the columns and test the solution at the half-way and the final stages with pH paper and a conductance bridge.

2 The softening of water

Titrate a 25 cm³ portion of tap-water with Wanklyn's soap solution (1 cm³ ≡ 1 mg calcium carbonate) until a permanent lather is obtained, i.e. one lasting up to three minutes. Also titrate a 25 cm³ portion of distilled water as a 'blank' titration.

Put about 25 cm³ of a cation exchange resin, in its sodium form, into a column and run tap-water through it at the rate of 5–10 cm³/min. After allowing 50 cm³ of water to run through the column titrate a 25 cm³ portion of the emergent solution with soap solution as before. Calculate the hardness of tap-water, in parts of calcium carbonate per million of water (mg/kg).

Alternatively the titrations may be done with edta (see experiment 6.46).

1.25 The Study of Sols

1 A sulphur sol prepared by precipitation and aggregation

Perform this experiment in a fume cupboard. Hydrogen sulphide (from a Kipp's apparatus) contains an impurity hydrogen chloride which can be removed by passing the gases through some water to rinse it. Pass the hydrogen sulphide next through $100 \, cm^3$ of distilled water in a beaker for a minute. Secondly pass sulphur dioxide from a siphon through $100 \, cm^3$ of distilled water in a beaker for a minute. Add the solution of sulphur dioxide to that of hydrogen sulphide until no black deposit forms on a piece of filter-paper soaked on lead acetate (ethanoate) solution which is held above the mixture.

Examine the colloidal solution that you have prepared. Can it be filtered either using filter-paper or Visking tubing? (Visking tubing is made of reconstituted cellulose; it has a pore size of about 2.4nm). Does a precipitate form on centrifuging? Does it show the Tyndall effect (the scattering of a beam of light converged by a lens)? Does it show Brownian movement (haphazard motion of the particles seen by using a microscope with a good depth of focus and preferably having illumination at right angles to the direction of observation)? Does it keep?

2 The coagulation of a sulphur sol

Divide the remainder of the sulphur sol into nine portions and add saturated, moderately concentrated and dilute solutions of each of sodium chloride, magnesium chloride and aluminium chloride to those portions. Observe the effects.

What is the Hardy-Schulze rule and is it followed here?

3 An iron(III) hydroxide sol prepared by hydrolysis and aggregation

To a spatula load of iron(III) chloride in a test-tube add $2 \, cm^3$ of distilled water: this gives a concentrated solution. Heat $1 \, cm^3$ of the solution until it boils: what else happens? Then heat a $250 \, cm^3$ beaker nearly full of distilled water until the temperature is about 370K (97°C); add one drop of the chloride solution from a teat pipette, then at intervals further drops until it is obvious that a sol has formed. The colloidal solution can be examined as in (1).

4 Dialysis of an iron(III) hydroxide sol

Suspend some of the iron(III) hydroxide sol in a U-tube made of Visking tubing in a beaker of distilled water. At intervals test samples of the water for the presence of chloride ions (by dilute nitric acid and silver nitrate solution) and of iron(III) ions (by potassium thiocyanate solution). pH paper may also reveal that dialysis has occurred.

5 Electrophoresis of an iron(III) hydroxide sol

Put some of the iron(III) hydroxide sol in the curved portion of a U-tube and with great care, pour distilled water into the two limbs simultaneously. Insert copper electrodes into the water and apply a potential of up to 24 volts: does this have any effect within about 30 minutes?

6 A silver chloride sol prepared by precipitation and aggregation

Another sol that can be easily prepared in the laboratory is silver chloride dispersed in Teepol or other liquid detergent. To $100 \, cm^3$ of the detergent add $10 \, cm^3$ of 0.1M hydrochloric acid and drip in, with stirring, 0.1M silver nitrate solution from a burette.

The sol is stabilized by the high surface tension of the detergent; it can be examined in the same way as above.

1.26 The Study of Gels and Emulsions

1 Silica gel prepared by hydrolysis

Dissolve about 2g of sodium silicate in 25 cm^3 of water, add two drops phenolphthalein and then run in 1 M hydrochloric acid from a burette until the pink colour has just been discharged.

Examine portions of the gel by heating and then cooling and by diluting with water; for the Tyndall effect; for Brownian movement; for elasticity (if slightly deformed does it regain its shape?); and fluidity (if an irregular lump is put in a test-tube does it soon conform to the shape of the bottom of the test-tube?).

2 Gelatine

Warm about 100 cm^3 of distilled water in a clean beaker and stir in about 2g of gelatine. Leave the solution to cool and then examine it as in (1).

3 The thixotropy of gels

Bentonite (and other clays), some paints, salad cream, tomato ketchup, etc. form gels which break down when subjected to mechanical agitation. Half-fill a test-tube with bentonite and add water to the same level. When the tube is shaken the mixture can be heard and felt to be moving but when the tube is inverted the mixture does not fall out.

4 Emulsions

An emulsion may be of water in oil or of oil in water. The addition of fluorescein which is water-soluble or of 'oil green' (or blue, etc. − oil soluble dyes) enables a distinction to be made.

Shake 2 cm^3 of water with 2 cm^3 of olive oil (alternatives are linseed oil, xylene and lubricating oil) for one minute and then note the time for them to separate. Examine the emulsion with fluorescein and oil green. Repeat each experiment in the presence of a few drops of soap solution.

Emulsions can be inverted: shake 3 cm^3 of xylene and 3 cm^3 of soap solution together, examine half with fluorescein and to the other half add a spatula load of magnesium sulphate crystals before re-examining with fluorescein.

Other emulsions which can be examined include hot and cold milk, hair cream, salad cream and butter.

1.27 Ascending, Descending and Radial Chromatography

Trim a piece of chromatography paper, e.g. a 20cm square, so that when smoothly coiled and clipped it will fit in a 1 dm³ beaker without touching the side or emerging from the top. Open the paper on a clean surface and rule a pencil line 2cm from the bottom. A spot of the unknown substance is put at the centre of this line; at about 2cm intervals on either side spots of the suspected constituents are placed using a capillary (melting-point) tube. The spots should be dried to stop them from spreading. Solvent is put in the beaker to a depth of 1 cm and the beaker is covered to reduce losses by evaporation. Slotted filter-paper may also be used.

If the apparatus is available cut a piece of chromatography paper, e.g. 25 cm of 10 cm wide paper, to fit in a tank for descending chromatography. Solvent to saturate the air in the tank should be placed in the bottom of the tank as well as in the trough at the top. The pencil line for the spots should be about 8 cm from one end of the strip of paper.

Next cut a wick from the edge to the centre of an 11 cm diameter piece of filter-paper and sandwich the paper between two sheets of glass. The wick should project through the hole in the centre of the lower sheet of glass into a beaker in which the solvent is placed. The unknown substance and the suspected constituents have to be put on separate pieces of paper in this method.

The experiment may be performed to compare three solvents, e.g. 85% acetone (propanone), 10% concentrated hydrochloric acid, 5% water; 60% butanol, 20% ethanol, 17.5% water, 2.5% concentrated hydrochloric acid; 50% diethyl ether (ethoxyethane), 30% methanol, 15% water, 5% concentrated hydrochloric acid, separating the same mixture by the same method. The experiment may also be used to compare the three methods of separating the same mixture by the same solvent.

Two mixtures can be examined in these ways: first, a universal indicator together with simple indicators, e.g. phenolphthalein, congo red, thymolphthalein, methyl red, methyl orange, etc, and second the substances studied in Group IV of the usual analysis tables, i.e. nickel, manganese, zinc and cobalt nitrates in solution (each about 10g/dm³).

The flow of solvent should be stopped before the front reaches the edge of the paper and the boundary marked in pencil. To make some of the indicators show up on the paper it may be necessary to fume it with ammonia issuing from concentrated ammonia solution or to spray it with 0.01 M ammonia solution and secondly to fume it with hydrogen chloride gas issuing from concentrated hydrochloric acid (or spray it with dilute acid). Any spots made visible should be ringed in pencil because they may fade later. To make some of the Group IV substances show up on the paper it may be sprayed with hydrogen sulphide solution or else with substances such as dimethylglyoxime (butanedione dioxime) etc. (see section 3.22). Dry the paper after spraying. Calculate the R_f values of the substances separately and in the mixture. If iron is present as an impurity it causes a green line near the solvent front.

$$R_f = \frac{\text{Distance moved by substance}}{\text{Distance moved by solvent}}$$

Compare and contrast the different methods in respect of density (compactness) of material after chromatographic separation, speed of separation, etc.

1.28 Electrophoresis and Paper Chromatography

1 Casein may be prepared from milk as follows. Put 10 cm³ of milk into a boiling tube and add 10 cm³ of water. Warm the tube in a beaker of water to 310 K (37°C) and maintain it at, or just below that temperature. Add 1.5 cm³ of 1M acetic (ethanoic) acid: this reduces the pH to below 4.6 and casein is precipitated during the next few minutes. Then add 1 cm³ of 1M sodium acetate solution which brings the pH up to 4.6. Stir the suspension and cool the tube under running water back to room temperature before adding an equal volume of water. Let the precipitate settle and decant off as much of the supernatant solution as possible. Wash the precipitate at least three times by stirring it with about 40 cm³ of water and decanting again.

2 Keep 0.02 g of casein for the chromatography experiment and to the remainder add 2g of urea (carbamide) and water to make the total volume 5 cm³. At the iso-electric point (pH = 4.6) the protein molecule will not migrate in an electric field and so the pH is raised to about 7 to cause it to move towards the anode. The buffer solution is made by dissolving 0.9g citric acid-1-water (2-hydroxypropane-1,2,3-tricarboxylic acid-1-water), 9 g disodium hydrogenphosphate — 12-water and 36 g of urea in water and making the total volume up to 100 cm³. The buffer solution is placed to exactly the same height in two 50 cm³ beakers and a glass rod held horizontally between the beakers and about 4cm above them: thus when a Λ-shaped strip of filter-paper is balanced on the glass rod it hangs with its ends in the buffer solutions. Carbon rods should be placed in the beakers and a potential of 60 volts applied.

The strip of filter-paper to be used must be clean and should be handled only with tweezers; the dispersion of casein in urea solution is put along the central line by means of a capillary tube or teat pipette. The assembly should be kept free from draughts by an inverted large beaker and the potential applied for about four hours. Then remove the strip from the beakers and dry it in a rapid stream of hot air. Soak the strip in 100 cm³ of methanol for five minutes, then in $50\,cm^3$ methanol containing $0.05\,g$ of bromophenol blue for 30 minutes and finally thrice for five minute periods in $200\,cm^3$ portions of 1 M acetic acid. Again dry the paper in a rapid stream of hot air. The various forms of casein are seen as yellow bands and they are made more evident by fuming with concentrated ammonia solution for a minute, which turns them blue.

---

3 The casein may also be hydrolyzed to find which simple amino acids it contains: boil about $0.02\,g$ of casein and $20\,cm^3$ of 6 M hydrochloric acid in a small flask fitted with a reflux condenser for up to one day (see figure 4.A). Then put the solution on a watchglass or in an evaporating basin held over a beaker of boiling water. The hydrolysate should then be put in a desiccator with soda-lime for one day. Dissolve the residue in a little warm water and evaporate the solution to dryness.

For the paper chromatography experiment dissolve the residue in $2\,cm^3$ of propan-2-ol containing four drops of water. Put a spot of the solution on a piece of chromatography paper as in experiment 1.27. Alongside the solution put spots of known aminoacids, e.g. lysine, histidine, glycine, serine, leucine, etc. The solvent employed is 60% butanol, 25% water, 15% acetic acid and the separation may take three hours or more. Then dry the paper before dipping it into ninhydrin [0.25% solution in acetone (propanone)]. The reaction between ninhydrin and the amino acids is caused by placing the paper in an oven set at about 350 K (77 °C) for about half-anhour. The bluish purple spots obtained should be ringed with a pencil because they may fade later. It is then possible to state some of the constituents of casein.

Ninhydrin

1.29 Thin Layer Chromatography

Either use a prepared strip of adsorbent or make your own: this is done by shaking $10\,g$ of Kieselgel G in $20\,cm^3$ of water for 1–2 minutes and then spreading it over 3–4 sheets of glass about 12 cm square so that a film about 0.3 mm thick is obtained — a glass rod functions as a rolling pin. After about 15 minutes put the plates in an oven set at about 360 K (87 °C) for 20 minutes.

Thin layer chromatography can be compared with ascending paper chromatography. Put the same pattern of spots, e.g. of indicators, on the paper and the thin layer chromatogram. The same solvent as in experiment 1.27 can be employed.

The results obtained by the two methods can be compared for speed of separation, compactness of spots, selectivity, etc.

1.30 Gas Chromatography

1 There are several instruments available now for experiments in gas chromatography. The column can be of 'White Tide' or of Carbowax 20M on fine granules (60–80 mesh) of an inert base. The detection device may be a thermocouple in the burning gas connected to an ammeter or a catharometer connected to a Wheatstone Bridge circuit. It is better to compare the carrier gas with the unknown mixture dispersed in the carrier gas than to examine the latter alone.

---

2 With only the carrier gas (natural gas or hydrogen) passing through the instrument a null reading should be obtained; it is advisable to dry the gas supplied with self-indicating silica gel. The materials which can be easily investigated are combustible gases and liquids with boiling-points below about 350 K (about 80 °C). A sample of the material under test is drawn into a glass syringe and then injected through a self-sealing rubber cap into the carrier gas. The time that elapses between injection and detection is recorded. If a mixture is injected then changes in the ammeter reading at several intervals will be recorded. A pen recorder may be used but otherwise the ammeter readings should be noted regularly and a graph plotted of current (μA) against time.

For a constant rate of flow of carrier gas the time interval between injection and detection, known as the retention time of a substance, is a constant. The area of the peak is a measure of the amount of the substance injected.

---

3 A simple experiment is to inject successively butane (from a camping stove), pentane, hexane and then a small sample of heptane. Plot a graph of the logarithm of the retention times against the thermodynamic boiling-points of these substances.

The effect of altering the amount of pentane upon the area of the peak may also be investigated.

1.31 Colorimetry: Beer's Law

1 Use a colorimeter. A light bulb does not emit all the colours of the spectrum at the same intensity, nor is the photo-conductive device (photoelectric-cell) uniformly sensitive. An ammeter can be used to assess the intensity of light: it may have a linear scale but the relationship is logarithmic. Beer's law (1852) states that

$$\log_{10} \frac{I_0}{I_t} = kc$$

where I_0 is a measure of the intensity of the incident light,

I_t is a measure of the intensity of the transmitted light,

c is the concentration of the solution in mol/dm^3

and k is a constant.

---

2 Put about $10\,cm^3$ of water in a clean test-tube: the meniscus of the water should be above the beam of light. Place the tube in the colorimeter and adjust the instrument to give the maximum meter reading possible. Then put each coloured filter in the colorimeter in turn and measure the respective currents (I_1). Repeat this procedure with the solution under test, e.g. copper(II) sulphate, and measure the respective currents (I_2) obtained. The best filter to use for an experiment is the one giving maximum absorption, i.e. the minimum value of I_2/I_1. This filter should be the one of the complementary hue to the colour of the solution, i.e. for the Ilford Bright Spectrum Filter Set:

Red   –   Green
Orange   –   Blue-green
Yellow   –   Blue
Yellow-green   –   Violet

---

3 Having chosen the correct filter for the solution of the substance to be examined put $10\,cm^3$ of water in the test-tube and with that filter in place adjust the instrument to give the maximum meter reading possible (I_3).

Then replace the water by $10\,cm^3$ of a concentrated solution of the substance under test, e.g. 1M copper sulphate, 0.02M potassium permanganate or 0.003M iodine, and measure the current (I_4). The solution must then be diluted accurately to obtain further values of I_4: using two burettes these volumes can be measured out (in cm^3).

Test solution 9 8 7 . . .
Water   1 2 3 . . .
and stirred thoroughly to ensure adequate mixing.

---

4 Then plot a graph of $\log_{10} I_4$ (vertical axis) against the concentration (horizontal axis) thus giving the calibration curve for that substance. The value of I_3 is constant and does not need to be taken into consideration for the graph. Beer's law applies with best results to dilute solutions and indicates that a linear relationship should be obtained.

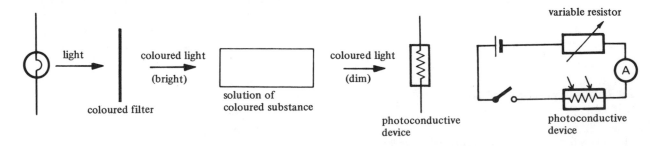

1.32 Colorimetry: The Study of Complex Formation

1 When small quantities of ammonia solution are added to copper(II) sulphate solution a pale blue precipitate of copper(II) hydroxide is obtained but if an excess is added a deep blue solution containing a copper-ammonia complex cation is given. To $1\,cm^3$ of $1M$ copper(II) sulphate add $9\,cm^3$ of $1M$ ammonia solution and stir the mixture well. Choose the appropriate filter for this deep blue solution as in experiment 1.31. Put $10\,cm^3$ of water in the test-tube and using this filter adjust the instrument to give the maximum meter reading possible (I_0).

2 Successively make up a set of mixtures in a small beaker from the component solutions and water (volumes in cm^3).

Copper(II) sulphate	2	2	2	2	2	2	2	2	2	2	2
Ammonia	5	6	7	8	9	10	11	12	13	14	15
Water	18	17	16	15	14	13	12	11	10	9	8

Stir each mixture well and decant some of the solution into a test-tube and centrifuge it well to precipitate any suspension; decant the clear solution into the test-tube for the instrument. Measure the currents obtained when these clear solutions are put into the instrument (I_t).

3 Then plot a graph of $\log_{10} I_t$ (vertical axis) against the volume of ammonia solution (horizontal axis). The value of I_0 is constant and does not need to be taken into consideration for the graph. Complex formation takes place in stages and the graph suggests the main stages.

1.33 Polarimetry: The Optical Activity of Some Substances

Use a polarimeter. A substance is optically active if it rotates the plane of polarization of plane polarized light. Although results can be obtained with white light, monochromatic light, e.g. that from sodium (at $589\,nm$) are better.

At a constant temperature the relationship of angle of rotation (θ) to concentration ($c\,g/cm^3$) and path lengths of light in the solution (l dm) is
$$\theta = \alpha l c$$

where α is the specific or optical rotation. The value of α must be divided by 100 if the path length is in metres and the concentration in g/dm^3 and it is desired to compare the answer obtained with that in a reference book.

Polaroid filters are used to polarize and analyze the light: when the analyzing filter (nearest the eye) is at $0°$ on the scale no light should pass through the instrument because the polarizing filter is at right angles to it. It is always advisable to check the zero of the instrument. When an optically active substance is placed between the filters the analyzing filter must be rotated to cut off the light again: the necessary angle of rotation may be an average (of the approaches clockwise and anti-clockwise) depending on the sensitivity of the instrument. The scale on the analyzing filter reads anti-clockwise because the direction of observation is the opposite of the direction of travel of the beam of light. Clockwise rotation of the beam of light is accorded a positive sign.

1 For a typical experiment 10g of maltose can be weighed out and made into $100\,cm^3$ of solution in a measuring flask. Maltose is a sugar, a disaccharide, with a high optical activity: it has the formula $C_{12}H_{22}O_{11}$ and is composed of two units of glucose less one of water. Put the solution into the instrument up to a predetermined level and measure the angle of rotation (θ). The solution does not deteriorate rapidly so $75\,cm^3$ of it can be diluted accurately to $100\,cm^3$ with distilled water and a new value of θ found for the more dilute solution. Successive dilutions yield more values of θ for the corresponding values of c. The path length of the solution in the instrument and the temperature should be recorded.

2 Plot a graph of θ (vertical axis) against c (horizontal axis) and measure the slope. The relationship of θ to c is linear and from the slope the value of α can be obtained.

3 An alternative experiment is to study the enantiotropes of an optically active compound in solutions of equal concentration, e.g. of tartaric (2,3–dihydroxy-butanedioic) acid.

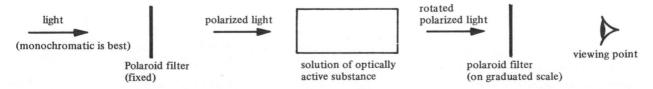

light → (monochromatic is best) | Polaroid filter (fixed) → polarized light | solution of optically active substance → rotated polarized light | polaroid filter (on graduated scale) → viewing point

1.34 The Depression of the Melting-Point (Beckmann)

1 A -10 to $+50^\circ$C thermometer is suitable. The melting-point of benzene (approximately 278 K, 5°C) is first found using this thermometer and then the depression caused by the addition of naphthalene.

2 By means of a pipette *and a filler* or a burette put 10cm^3 of benzene in a clean dry test-tube and dip this directly into the freezing mixture until the benzene just starts to freeze. Then warming the test-tube with your hand will just melt the benzene and the test-tube can be inserted in its air-jacket (the boiling tube) which will steady any changes in temperature. **Care:** use a fume cupboard.

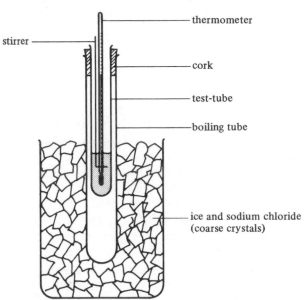

thermometer

stirrer

cork

test-tube

boiling tube

ice and sodium chloride (coarse crystals)

The temperature should be allowed to fall below the melting-point, i.e. supercooling allowed, before stirring. As the enthalpy of melting (latent heat of fusion) is evolved the temperature rises, and is maintained at a steady value during solidification, before falling again. The steady temperature, according to your thermometer, is the melting-point of benzene.

3 To the benzene add about 0.3g of naphthalene, weighed out accurately. The melting-point of the solution is found by allowing the solution to cool in the air-jacket, stirring all the time, i.e. no supercooling is permitted because it is essential that crystals of the solvent separate. The experiment may be continued if there is time by adding further portions of naphthalene up to a maximum of 1.2g.

4 The density of benzene is 0.88 g/cm^3. It is found that 1 mole of a substance dissolving in 1 kg of benzene depresses the melting-point by 5.12K ($^\circ$C). If only one result has been obtained the relative molecular mass of naphthalene should be calculated in accordance with this statement.

If several results have been obtained a graph of the melting-point (vertical axis) against the mass of substance (horizontal axis) should be plotted: the graph should be a straight line and the melting-point of benzene is a point on the graph. From the slope of the line the relative molecular mass of naphthalene can be calculated. The answer obtained should be accurate enough to decide what multiple of the relative empirical mass to take. What value of M_r do you obtain?

1.35 The Depression of the Melting-Point (Rast)

1 In Rast's method a 0–360°C thermometer is used because the cryoscopic constant for camphor is large and so are the depressions in melting-point. One end of each of the melting-point tubes should be sealed by fusion in a flame: the tubes can only be used once. A melting-point bath of medicinal paraffin or of silicone oil (MS 550) can be used.

2 Powder some camphor in a clean dry mortar with a pestle and put some in a melting-point tube to a depth of 1 cm: this is a tricky operation — vibration or tapping may help and a clean copper wire may be used as well. The melting-point tube should adhere by surface tension to the thermometer in the bath — if not a rubber band may be used — and the sample should be at the same level as the bulb of the thermometer. The bath should be heated with a burner, a micro-burner being advantageous, and the source of heating removed as the melting-point is approached: pure camphor melts at approximately 449 K (176°C). At the melting-point the powder in the tube goes from cloudy to clear and may sag in the tube. The melting-point of the camphor used according to your thermometer should be found twice and the average taken.

3 Into a 100×16mm test-tube weigh accurately about 0.1g of naphthalene and 1g of camphor. **Just** melt the contents of the tube **rapidly** in a **small** hot Bunsen flame: camphor is volatile and care must be exercised so that as little as possible sublimes away. With a clean dry glass rod thoroughly stir the mixture and when it has resolidified scrape it out into the mortar and grind it thoroughly. Find the melting-point of the solution as above.

4 It is found that 1 mole of a substance dissolving in 1 kg of camphor depresses the melting-point by 40 K ($^\circ$C). Calculate the relative molecular mass of naphthalene in accordance with this statement. The answer obtained should be accurate enough to decide what multiple of the relative empirical mass to take. What value of M_r do you obtain?

1.36 The Elevation of the Boiling-Point (Beckmann)

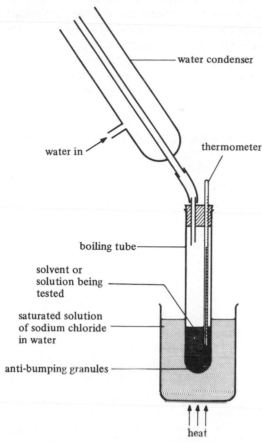

1 A 50–105°C thermometer is suitable. Into the boiling-tube put 25 cm³ of distilled water and using sodium chloride solution as a heating bath heat the water and determine its boiling-point.

2 Allow the apparatus to cool for a moment before disconnecting the condenser and pouring in approximately 2g of anhydrous glucose which has been weighed out accurately. Determine the boiling-point of the solution as before. The experiment may be continued if there is time by adding further portions of glucose up to a maximum of 8g.

3 It is found that 1 mole of a substance dissolved in 1 kg of water raises the boiling-point by 0.52K (°C). If only one result has been obtained the relative molecular mass of glucose should be calculated in accordance with this statement.

   If several results have been obtained a graph of the boiling-point (vertical axis) against the mass of substance (horizontal axis) should be plotted: the graph is a straight line and the boiling-point of water is a point on the graph. From the slope of the line the relative molecular mass of glucose can be calculated. The answer obtained should be accurate enough to decide what multiple of the relative empirical mass to take. What value of M_r do you obtain?

1.37 The Elevation of the Boiling-Point (Landsberger)

1 Hot water may be used to fill the steam generator so that the experiment is accelerated. Steam is generated and passed through the tube until a steady temperature is reached: a 50–105°C thermometer is suitable. The reading obtained on the thermometer is the boiling-point of water at the prevailing atmospheric pressure.

2 Then tip out any condensed steam and weigh in accurately about 1.5 g of urea (carbamide). This can be put in using a dry funnel so that it falls directly to the bottom of the inner tube: if you fail to accomplish this use a minute quantity of hot distilled water to wash the crystals down. Pass in steam again. The moment all the crystals have dissolved disconnect the steam generator and rapidly read the temperature and the volume of the solution. This will be the maximum temperature recorded: why? Also why does steam at 373K (100°C) cause the solution to boil at about 378 K (105 °C)?

3 Pass in more steam, disconnecting the supply at intervals to record the temperature and the volume. Finally dismantle the apparatus carefully noting the position of the thermometer and steam tube in the inner tube. Empty out the solution. Run in water from a burette and at intervals note the burette reading and, with the steam tube and thermometer in, the level in the inner tube. This will enable you to correct the experimental volumes to true volumes.

4 It is found that 1 mole of a substance dissolved in 1 kg of water raises the boiling-point by 0.52K (°C). Plot a graph of the boiling-point (vertical axis) against the reciprocal of the volume (horizontal axis); the graph is a straight line, the boiling point of water being a point on the graph. From the slope of the line the relative molecular mass of urea can be calculated. The answer obtained should be accurate enough to decide what multiple of the relative empirical mass to take. What value of M_r do you obtain?

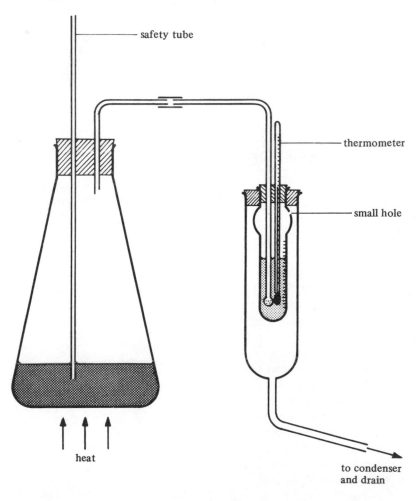

safety tube

thermometer

small hole

heat

to condenser and drain

1.38 Electrolytes and the Elevation of the Boiling-Point

1 Clamp a $100\,cm^3$ round-bottomed flask containing several anti-bumping granules above a tripod and gauze. Have an accurate thermometer [reading up to $383\,K\ (110\,°C)$] fitted, but not corked, into the flask so that it reaches almost to the bottom. By pipette, put in $10\,cm^3$ of distilled water and measure its boiling-point when it is heated rapidly. Pour away the water and allow the flask to drain.

---

2 Weigh out accurately about $45\,g$ of fructose ($C_6H_{12}O_6$, 0.25 moles, relative molecular mass 180) and make it into $50\,cm^3$ of solution with distilled water. By pipette put $10\,cm^3$ of solution into the flask, add several anti-bumping granules, heat rapidly and observe the boiling-point. Pour away the solution, rinse the flask with hot distilled water and allow it to drain. Repeat the experiment and take an average of the boiling-points observed. Calculate the elevation of the boiling-point (the ebullioscopic constant) which would be caused if there were 1 mole of fructose in $1\,dm^3$ of solution.

---

3 Next, weigh out accurately about $14\,g$ of sodium chloride (NaCl, 0.25 moles, relative molecular mass 58.5) and make it into $50\,cm^3$ of solution with distilled water. By pipette put $10\,cm^3$ of the salt solution into the flask, add several anti-bumping granules, heat rapidly and observe the boiling-point. Pour away the solution, rinse the flask with hot distilled water and allow it to drain. Repeat the experiment and take an average of the boiling-points observed.

---

4 Calculate the elevation of the boiling-point which would be caused if there were 1 mole of sodium chloride in $1\,dm^3$ of solution.

   Calculate the apparent degree of dissociation of the sodium chloride in solution at about 373 K ($100\,°C$) using the relationships:

   van't Hoff factor

   $$i = \frac{\text{Elevation of boiling-point for 1M NaCl solution}}{\text{Elevation of boiling-point for 1M fructose solution}}$$

   Arrhenius, degree of dissociation

   $$\alpha = \frac{i-1}{n-1}$$

where n is the number of ions from one empirical formula of the sodium chloride when it dissociates, and α is the degree of dissociation.

The Rates of Reactions

1.39 The Catalytic Decomposition of Hydrogen Peroxide

$$2H_2O_2 \rightarrow 2H_2O + O_2 \uparrow$$

1 Set up the apparatus as shown. The piston of the syringe should be moved slowly by a corkscrew action. The water in the manometer will indicate whether or not the pressure inside the apparatus is the same as that outside. If the piston is moved slightly and the water level does not move to a new fixed position then the apparatus must be investigated for a leak. The second conical flask is to steady out any changes in the pressure inside the apparatus.

2 Put $100\,cm^3$ of water by measuring cylinder into the conical flask and then sprinkle in about $0.25\,g$ of manganese(IV) oxide. If each person in the class weighs out accurately a different mass of this catalyst then it is possible to determine whether the quantity of catalyst has any influence upon the rate of reaction. Suspend a small test-tube (up to $75 \times 10\,mm$) vertically inside the conical flask by means of a thread, having in the tube $2\,cm^3$ of 50-volume hydrogen peroxide solution.

3 Adjust the piston so that the pressure inside the apparatus is the same as outside and then release the thread at a known time (designated zero for this experiment). Gently oscillate the flask for the rest of the experimental time. As oxygen is released move the piston to the right to keep the pressure constant and record the volume at times such as 1, 2, 3, 4, 5, 10, 15, 20 . . . minutes until the volume remains constant (it should reach $100\,cm^3$). Record room temperature.

4 Let V_t be the volume of oxygen at time t and V_∞ be the final volume. Plot a graph of the oxygen released (vertical axis) against the time (horizontal axis). Then plot a graph of $\log_{10} V_\infty/(V_\infty - V_t)$ (vertical axis) against time (in seconds, horizontal axis).

The slope of the second graph, which should be a straight line because it is a reaction of the first order, is $k/2.303$, where k is the velocity constant. The units of k are s^{-1}.

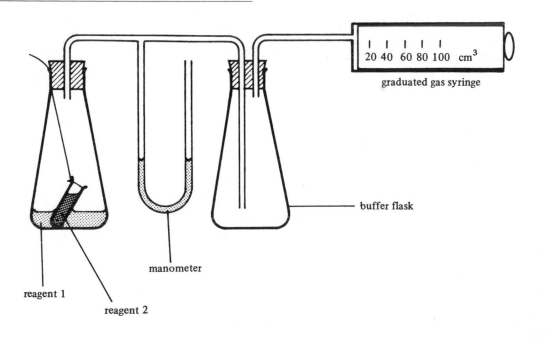

graduated gas syringe

buffer flask

manometer

reagent 1

reagent 2

1.40 The Reaction of Nitric Acid with Sodium Thiosulphate [Thiosulphate (VI)]

$$2H^+ + S_2O_3^{2-} \rightarrow S\downarrow + H_2O + SO_2$$

1 The sulphur dioxide remains in solution at room temperature. The precipitation of sulphur causes the solution to become opaque; so the rate of reaction can be studied by measuring the time taken (in seconds) to cause enough precipitation to obscure writing on a piece of paper below the reaction vessel. In each experiment the total volume of solution is 20cm³ so that the volumes of the individual solutions are directly proportional to their concentrations.

2 Using three burettes put the sodium thiosulphate solution and the water in a small beaker and the nitric acid in a test-tube (volumes in cm³).

Experiment	Sodium thio-sulphate (1M)	Water	Nitric Acid (0.1M)
A	2	8	10
B	4	6	10
C	6	4	10
D	8	2	10
E	10	0	10

At a convenient time add the acid to the sodium thiosulphate solution and record the time for the reaction. Record room temperature.

3 The experiment may be continued by investigating the effect of changing the concentration of acid: use 6 cm³ 1M sodium thiosulphate solution mixed with 6, 8, 10, 12, 14cm³ of the nitric acid (and 8, 6, 4, 2, 0cm³ water respectively).

   Alternatively the experiment may be continued by investigating the rate for 6 cm³ of the sodium thiosulphate mixed with 6 cm³ of the nitric acid and 8 cm³ water at 10 K (°C) intervals above room temperature.

4 The rate of reaction is assessed by the reciprocal of the time taken to produce obscurity. Plot a graph of t^{-1} (vertical axis) against the volume of sodium thiosulphate taken (horizontal axis); the origin is a point on the graph. The graph should be a straight line proving that the reaction is of the first order with respect to the sodium thiosulphate.

   What is the order of the reaction with respect to the concentration of the acid?

   If the temperature is varied see experiment 1.48 for the treatment of the results.

1.41 The Use of a Colorimeter to Study a Precipitation Reaction

This experiment depends on the fact that light is scattered by colloidal sulphur and hence as the reaction proceeds the intensity of the light reaching the photo-electric cell decreases. A filter is not needed. Put 10cm³ of water in a test-tube and place the tube in the colorimeter; adjust the instrument to give the maximum meter reading possible. See also experiment 1.31.

   The rate of the reaction is assessed by finding the time for the meter reading to fall to a convenient value (about half to two-thirds of the initial reading) for various mixtures of 0.05 M sodium thiosulphate [thiosulphate(VI)] solution and water, 1 cm³ of 0.1 M hydrochloric acid being stirred in at zero time.

Sodium thiosulphate (cm³)	9 8 7 6 5
Water (cm³)	0 1 2 3 4

Record the temperature of the room.

   The rate of the reaction is assessed by the reciprocal of the time taken for the change in the meter reading. Plot a graph of the rate of the reaction (vertical axis) against the concentration of the sodium thiosulphate (assessed by the volume taken; horizontal axis). The graph should be a straight line showing that the reaction is of the first order with respect to the sodium thiosulphate.

1.42 The Reaction of a Sulphite [Sulphate(IV)] with an Iodate [Iodate(V)]

The reaction

$$IO_3^- + 3SO_3^{2-} \rightarrow I^- + 3SO_4^{2-}$$

is slow even though it is catalyzed by acids but it is followed by a fast reaction

$$5I^- + IO_3^- + 6H^+ \rightarrow 3I_2 + 3H_2O$$

The iodine released gives a blue coloration with starch: a tunnel clathrate. This experiment can be done using measuring cylinders, each member of the class being responsible for one section and the results then collated. The total volume of solutions is $400\,cm^3$ in each part.

Expt	Potassium iodate (0.05M)	H_2O	Sodium sulphite (0.025M)	H_2O	Sulphuric acid (0.25M)	H_2O	Starch 4 g/dm^3
1	8	192	25	75	10	80	10
2	10	190	25	75	10	80	10
3	12	188	25	75	10	80	10
4	14	186	25	75	10	80	10
5	16	184	25	75	10	80	10
6	20	180	2	98	10	80	10
7	20	180	4	96	10	80	10
8	20	180	6	94	10	80	10
9	20	180	8	92	10	80	10
10	20	180	10	90	10	80	10
11	20	180	25	75	2	88	10
12	20	180	25	75	4	86	10
13	20	180	25	75	6	84	10
14	20	180	25	75	8	82	10
15	20	180	25	75	10	80	10

By a small measuring cylinder put the iodate solution into a large measuring cylinder and then make up the volume to $200\,cm^3$. By a second small measuring cylinder put the sulphite solution into a second large measuring cylinder and make up the volume to $100\,cm^3$. From a third small measuring cylinder add the acid to the sulphite and make up the volume to $190\,cm^3$. Finally from a fourth small measuring cylinder add the starch solution to the sulphite and acid thus making up the volume to $200\,cm^3$.

At zero time pour the contents of both large measuring cylinders simultaneously into a large beaker and stir well. Record the time taken for the appearance of a blue coloration. Note the temperature of the room.

Experiments (1) to (5) show the effect of varying the iodate concentration upon the rate of the reaction, the other factors being kept constant. Plot a graph of the rate of the reaction assessed by the reciprocal of the time taken (vertical axis) against the concentration of the iodate assessed by the volume taken (horizontal axis). The origin is a point on the graph.

Experiments (6) to (10) show the effect of varying the sulphite concentration and are dealt with as shown. Similarly experiments (11) to (15) show the effect of varying the acid concentration. If suitable equations can be found to fit the graphs obtained the order of the reaction with respect to each component can be found and hence the total order of the reaction.

1.43 To Determine the Velocity Constant of a Hydrolysis

Methyl formate (methanoate) in an acidic solution hydrolyzes quite rapidly and the extent of the reaction can be assessed by titration with standard alkali. The hydrolysis should be carried out at a constant temperature using a thermostatically controlled water-bath.

A 'control' titration must be done first: take $2\,cm^3$ of hydrochloric acid (approximately 0.5M) by pipette and run it into a $100\,cm^3$ of distilled water (measured out in a measuring cylinder and put in a conical flask). This procedure has to be adopted in subsequent experiments to stop the hydrolysis continuing whilst the titration is done and is adopted here so that all titrations are done under the same conditions. Titrate the acid with sodium hydroxide solution (approximately 0.1M) using phenolphthalein as the indicator. This titre measures the quantity of alkali neutralized by the catalyst and later titrations are greater because formic (methanoic) acid is also present in the samples taken.

$$HCOOCH_3 + H_2O \xrightarrow{H^+} HCOOH + CH_3OH$$

At zero time pour about $10\,cm^3$ of methyl formate into approximately $200\,cm^3$ of the hydrochloric acid in a conical flask in a water-bath: set it at 298K (25°C). At intervals withdraw $2\,cm^3$ samples from the conical flask and run them, separately, into $100\,cm^3$ portions of distilled water as above. The times of dilution should be for example 1, 5, 10, 20, 30, 40, 50, 60, 120 minutes and about four hours (or until the last two titres are identical). Titrate each sample with alkali as before.

Let the titre initially be $V_o\,cm^3$, at time t be $V_t\,cm^3$ and finally be $V_\infty\,cm^3$. The concentration of formic acid at a given time is proportional to $V_t - V_o\,cm^3$. Plot a graph of $V_t - V_o$ (vertical axis) against t (horizontal axis). Comment upon the shape of this graph. Secondly plot a graph of $\log_{10}(V_\infty - V_o)/(V_\infty - V_t)$ (vertical axis) against t (horizontal axis). If the second graph is a straight line it proves that the reaction is of the first order. The slope of the second graph is $k/2.303$, where k is the velocity constant. The units of k are s^{-1}. The reaction is however, bimolecular: explain this term.

1.44 The Use of a Polarimeter to Study a Hydrolysis

A polarimeter can be used to study the rate of hydrolysis of an optically active substance, e.g. sucrose, when catalyzed by a dilute solution of a strong acid. See also experiment 1.33.

Weigh out about 10g of sucrose (cane sugar) and stir it with water until $50\,cm^3$ of solution are obtained. Measure the angle of rotation in the polarimeter for the light to be cut off (θ_o).

Weigh out the same amount of sucrose again. Rapidly shake the sugar with $25\,cm^3$ of water, stir in $15\,cm^3$ of 2M hydrochloric acid and add water to make the total volume $50\,cm^3$, if that is the volume of the solution to put in the polarimeter tube. Zero time is the time at which the acid is added because in its absence the rate of hydrolysis is negligible. At times such as 5, 10, 15, 20, 25, 30, 45 and 60 minutes obtain a reading on the polarimeter. In between readings the tube should be kept in a thermostatically controlled water-bath set at 298K (25°C). Constant readings mark the end of the hydrolysis and this may be confirmed about 24 hours later.

$$\underset{\substack{\text{Sucrose} \\ \text{(dextra-rotatory)}}}{C_{12}H_{22}O_{11}} + H_2O \xrightarrow{H^+} \underset{\substack{\text{Glucose} \\ \text{(dextra-} \\ \text{rotatory)}}}{C_6H_{12}O_6} + \underset{\substack{\text{Fructose} \\ \text{(laevo-rotatory)}}}{C_6H_{12}O_6}$$

Let the reading at time t be θ_t and finally θ_∞. The concentration of sucrose is measured by $\theta_\infty - \theta_t$. As the experiment proceeds plot a graph of $\theta_t - \theta_o$ (vertical axis) against t (horizontal axis). Comment upon the shape of this graph.

Secondly plot a graph of $\log_{10}(\theta_\infty - \theta_o)/(\theta_\infty - \theta_t)$ (vertical axis) against t (horizontal axis). If the second graph is a straight line it proves that the reaction is of the first order. The slope of the second graph is $k/2.303$, where k is the velocity constant. The units of k are s^{-1}. The reaction is, however, bimolecular: explain this term.

1.45 The Reaction of Hydrogen Peroxide with an Iodide

$$H_2O_2 + 2I^- + 2H^+ \rightarrow I_2 + 2H_2O$$
$$I_2 + 2S_2O_3^{2-} \rightarrow 2I^- + S_4O_6^{2-}$$

1 The rate of oxidation of potassium iodide is found by measuring the time taken to cause enough iodine to be liberated to react with $1\,cm^3$ $0.01\,M$ sodium thiosulphate(VI) solution and then to give a blue colour with starch. The coloration must be intense enough to obscure writing on a piece of paper below the reaction vessel. Starch with iodine gives a tunnel clathrate (see introduction to experiment 6.29). Alternatively a colorimeter may be used to time the reaction, compare experiment 1.41.

2 By pipettes and burettes put the following solutions in a small beaker (volumes in cm^3):

Expt	Sulphuric acid (0.05M)	Starch solution (4 g/dm^3)	Potassium iodide (0.5M)	Water	Sodium thiosulphate (0.01M)
1	25	1	1	4	1
2	25	1	2	3	1
3	25	1	3	2	1
4	25	1	4	1	1
5	25	1	5	0	1

Into a small test-tube put $1\,cm^3$ of $0.25\,M$ hydrogen peroxide and at a convenient time add this to the mixture in the beaker: record the time for the reaction. Note the temperature of the room.

The rate of reaction is assessed by the reciprocal of the time; the concentration of the iodide is assessed by the volume taken because the total volume is constant. Plot a graph of reciprocal time (vertical axis) against the volume of iodide solution (horizontal axis). A straight line proves that the reaction is of the first order with respect to the iodide concentration.

3 To find the dependence of the rate of the reaction upon the concentration of peroxide, put the following solutions by pipettes and burettes in a small beaker (volumes in cm^3):

Expt	Sulphuric acid (0.05M)	Starch solution (4 g/dm^3)	Hydrogen peroxide (0.25M)	Water	Sodium thiosulphate (0.01M)
6	25	1	1	4	1
7	25	1	2	3	1
8	25	1	3	2	1
9	25	1	4	1	1
10	25	1	5	0	1

Into a small test-tube put $1\,cm^3$ of $0.5\,M$ potassium iodide solution and proceed as in (2).

Plot a graph of reciprocal time (vertical axis) against the volume of hydrogen peroxide (horizontal axis).

4 To find the dependence of the rate of the reaction upon the concentration of acid proceed as in step (2) but reduce the volume of acid $25, 20, 15, 10, 5\,cm^3$ and increase the volume of water to $4, 9, 14, 19, 24$ cm^3 respectively whilst keeping the other volumes constant at $1\,cm^3$. The reaction proceeds in more than one stage and by more than one path so when a graph of reciprocal time (vertical axis) is plotted against the volume of acid used (horizontal axis) a straight line is obtained which does not pass through the origin.

5 The reaction may be studied at different temperatures in order that the energy of activation can be calculated (as in experiment 1.48). Ammonium molybdate solution ($1\,g/dm^3$) exerts a catalytic effect and this also may be studied. This is the Harcourt-Esson reaction (1867); it is sometimes called a 'clock' reaction.

1.46 The Reaction of a Bromate [Bromate(V)] with an Iodide

$$BrO_3^- + 6H^+ + 6I^- \rightarrow 3I_2 + Br^- + 3H_2O$$

1 The rate of oxidation of potassium iodide is found by measuring the time taken to cause enough iodine to be liberated to react with $1\,cm^3$ 0.01M sodium thiosulphate(VI) solution as in the previous experiment, alternatively a colorimeter may be used or thirdly a control solution may be made up and the experimental solution compared. A suitable control solution is made by mixing one drop $(0.05\,cm^3)$ of 0.02M potassium iodide solution with $0.5\,cm^3$ of 0.5M hydrochloric acid and $0.5\,cm^3$ of 0.25 M hydrogen peroxide and then adding $5\,cm^3$ of the starch solution and $24\,cm^3$ of water.

2 By burettes put the following solutions into a small beaker (volumes in cm^3):

Expt	Potassium bromate (0.02M)	Hydrochloric acid (0.5M)	Water	Starch (4g/dm^3)	Sodium thiosulphate (0.01M)
1	12	2	4	5	1
2	10	2	6	5	1
3	8	2	8	5	1
4	6	2	10	5	1
5	4	2	12	5	1

Into a test-tube put $6\,cm^3$ of 0.02M potassium iodide solution and at a convenient time add this to the mixture in the beaker. Record the time taken for the blue colour to appear.

The rate of reaction is assessed by the reciprocal of the time; the concentration of the bromate is assessed by the volume taken because the total volume is constant. Plot a graph of reciprocal time (vertical axis) against the volume of bromate solution (horizontal axis). What is the order of the reaction with respect to the bromate concentration? Record the temperature of the room.

3 To find the dependence of the rate of the reaction upon the concentration of the iodide, put the following solutions by burettes into a small beaker (volumes in cm^3):

Expt	Potassium iodide (0.02M)	Hydrochloric acid (0.5M)	Water	Starch (4g/dm^3)	Sodium thiosulphate (0.01M)
6	12	2	4	5	1
7	10	2	6	5	1
8	8	2	8	5	1
9	6	2	10	5	1
10	4	2	12	5	1

Into a test-tube put $6\,cm^3$ of 0.02M potassium bromate solution and proceed as in (2). Plot a graph of reciprocal time (vertical axis) against the volume of potassium iodide solution (horizontal axis).

4 Is it likely that the 13 ions shown in the equation meet simultaneously to cause the reaction? In the light of your results suggest a possible mechanism for the slowest step in the reaction.

1.47 The Reaction of a Persulphate [Peroxo-disulphate(VI)] with an Iodide

$$S_2O_8^{2-} + 2I^- \rightarrow 2SO_4^{2-} + I_2$$

1 Put $25\,cm^3$ of $0.5\,M$ potassium iodide solution and $10\,cm^3$ of $0.02\,M$ potassium persulphate solution into a conical flask and heat the mixture to about 330K (about 57°C) and maintain it at that temperature for five minutes to complete the reaction. Titrate the mixture with $0.025\,M$ sodium thiosulphate(VI) solution (let the titre be $V_\infty\,cm^3$).

2 The rate of the reaction at room temperature can be followed by adding a known volume ($V_t\,cm^3$) of sodium thiosulphate solution and measuring the time taken for an equivalent amount of iodine to be produced, after which the familiar blue colour appears.

Into one conical flask put $25\,cm^3$ of the potassium iodide solution, $10\,cm^3$ of water and about $1\,cm^3$ of starch solution. Into a second flask put $10\,cm^3$ of the potassium persulphate solution and $25\,cm^3$ of water. Fill a burette up to the zero mark with the sodium thiosulphate solution and add $5\,cm^3$ to the second flask. At a convenient time pour the contents of the first flask into the second flask and swirl the contents to ensure thorough mixing. When the blue coloration appears record the time and then add a further $2\,cm^3$ from the burette. Continue the addition of solution in $1\,cm^3$ portions and note the time on every occasion the blue coloration appears: this may be continued until about $15\,cm^3$ have been added. Record room temperature.

3 The difference between V_∞ and V_t is a measure of the concentration of potassium persulphate at time t. Plot a graph of $V_\infty - V_t$ (vertical axis) against t (horizontal axis) and comment upon its shape. Secondly plot a graph of $\log_{10} V_\infty/(V_\infty - V_t)$ (vertical axis) against t (horizontal axis) and comment upon its shape. Is the reaction of the first order with respect to potassium persulphate?

1.48 To Determine the Energy of Activation of a Reaction

$$2MnO_4^- + 16H^+ + 5C_2O_4^{2-} \rightarrow 2Mn^{2+} + 8H_2O + 10CO_2\uparrow$$
Purple   Colourless
                                          in solution

1 The rate of the reaction at various temperatures of permanganate [manganate(VII)] ions with oxalate (ethanedioate) ions is investigated by finding the time taken for the purple colour to disappear after mixing solutions of these ions. Any brown suspension is disregarded. The oxalic (ethanedioic) acid solution ($0.05\,M$) is made up in $1\,M$ sulphuric acid.

2 By pipette (**N.B.** use a filler) put $10\,cm^3$ potassium permanganate ($0.02\,M$) and $10\,cm^3$ of the oxalic acid into separate boiling tubes. Put the boiling tubes into a beaker of water and heat them to about 320K (about 47°C). Remove the burner from under the tripod and allow the temperature to become steady before pouring one solution into the other. Stir the solution with the thermometer and record the time (in seconds) for the reaction. The average of the temperatures at the start and the finish of the reaction is taken as the temperature of the reaction.

3 Repeat (2) several times stopping the heating of the beaker at 330K (57°C), 340K (67°C), 350K (77°C) and 360K (87°C).

4 Plot a graph of the time of the reaction (vertical axis) against the temperature (horizontal axis). Does raising the temperature e.g. from 340 to 350K (67 to 77°C), halve the time of the reaction?

The rate of the reaction is assessed by the reciprocal of the time. Secondly a graph can be plotted of the logarithm to the base 10 of the time (vertical axis) against the reciprocal of the thermodynamic temperature (horizontal axis).

From the Arrhenius equation ($k = Ae^{-E/RT}$) it can be shown that the slope of the graph is

$$\frac{E}{2.303 \times 8.314}$$

where E is the activation energy of the reaction in joule/mole. Calculate E.

1.49 A Study of the Iodination of Acetone (Propanone)

$$I_2 + CH_3COCH_3 \rightarrow CH_3COCH_2I + H^+ + I^-$$
$$H^+ + HCO_3^- \rightarrow H_2O + CO_2 \uparrow$$
$$I_2 + 2S_2O_3^{2-} \rightarrow 2I^- + S_4O_6^{2-}$$

1 The acid catalyzed iodination of acetone in aqueous solution proceeds at a rate which is directly proportional to the concentration of acetone. The dependence of the rate upon the concentration of iodine may be determined in experiments in which such a large excess of acetone is present that its concentration is virtually unchanged throughout the reaction. The progress of the reaction is followed by withdrawing a portion of the reaction mixture and pouring it into a sodium hydrogencarbonate solution to neutralize the hydrogen ions liberated before titrating with sodium thiosulphate(VI) using starch as an indicator.

2 Into one conical flask put $50 cm^3$ of $0.02 M$ iodine dissolved in $0.2 M$ potassium iodide solution. Into a second flask put $25 cm^3$ of a $1 M$ solution of acetone in water and $25 cm^3$ of $1 M$ sulphuric acid. Have several other flasks (at least three) available, and put into them $10 cm^3$ portions of $0.5 M$ sodium hydrogencarbonate solution.

3 At a convenient time pour the contents of the first flask into the second flask and shake well for at least one minute. Withdraw a $10 cm^3$ portion of the reaction mixture and pour it into $10 cm^3$ of $0.5 M$ sodium hydrogencarbonate solution in the third flask: this should be done at times such as 1, 5, 10, 15, 20, 30, 40, 50 and 60 (maximum) minutes.

   When the bubbling has ceased titrate the residual iodine with $0.01 M$ sodium thiosulphate solution using starch as an indicator towards the end of the titration. It is advisable to rinse out all flasks immediately after use. Record the temperature of the room.

4 Plot a graph of the titres (vertical axis) against the time (horizontal axis). Comment upon the shape of the graph and deduce the order of the reaction with respect to the iodine.

1.50 The Study of a Hydrolysis by Conductance

1 Check the operation of the conductance bridge using a known resistor (see the paragraphs above experiment 1.4). Add $10 cm^3$ of distilled water to $40 cm^3$ of ethanol (methylated spirits) and put this aqueous ethanol in a $100 cm^3$ beaker and stir well (use a magnetic stirrer, if available).

$$(CH_3)_3CBr + H_2O \rightarrow (CH_3)_3COH + H^+ + Br^-$$

2 At a convenient time add $0.3 cm^3$ of *tertiary* butyl bromide (2-bromo-2-methylpropane) to the aqueous ethanol. At times such as 3, 6 . . . 30, and then 40, 50 and 60 minutes measure the conductance of the solution (S_t, in siemens or ohm^{-1}). A final reading (S_∞) should be obtained at least six hours, possibly 24 hours, later. Record the temperature of the solution.

3 Plot a graph of S_t (vertical axis) against time (horizontal axis) and comment upon its shape. Secondly plot a graph of $\log_{10} S_\infty/(S_\infty - S_t)$ (vertical axis) against time (horizontal axis). If a straight line is obtained the reaction is of the first order with respect to the halide concentration and the velocity constant may be calculated from the slope ($k/2.303$, in s^{-1}).

1.51 The Intermediate States of Reactions

Reactions rarely proceed directly from the reagents to the products but in many cases this is hard to demonstrate. Here are two examples which by their colours show the presence of intermediate states.

1 **A chromium compound**
Add hydrogen peroxide to a solution of potassium dichromate [dichromate(VI), an orange colour] acidified with dilute sulphuric acid. A blue colour appears and finally a green coloured solution containing a chromium(III) salt is left. The blue coloured substance is soluble in and is stabilized by diethyl ether (ethoxyethane). **Care:** no flames must be nearby. The blue colour is caused by the intermediate compound chromium pentaoxide [chromium(VI) oxide diperoxide].

$$Cr_2O_7^{2-} + 4H_2O_2 + 2H^+ \rightarrow 2CrO_5 + 5H_2O$$
$$2CrO_5 + 6H^+ \rightarrow 2Cr^{3+} + 2H_2O + H_2O_2 + 3O_2 \uparrow$$

The experiment may be started with potassium chromate [chromate(VI), yellow] which becomes the dichromate upon acidification.

2 **A cobalt compound**
Hydrogen peroxide (20-volume) is not decomposed even if it is boiled with an excess of a $0.5 M$ solution of potassium sodium tartrate (Rochelle salt). Demonstrate that this is so.

   Next, to the hot (but not boiling) solution add cobalt(II) chloride solution until the mixture is pink. There is a pause, then the solution turns green while oxygen is evolved and finally the pink colour returns.

Reversible Reactions

1.52 Some Qualitative Examples

1 Copper(II) sulphate

Stir a spatula load of copper(II) sulphate-5-water crystals into $3\,cm^3$ of concentrated sulphuric acid. The crystals lose their colour but it is regained if the solution is poured into $6\,cm^3$ of water (in a small beaker).

$$CuSO_4 \cdot 5H_2O \rightleftharpoons CuSO_4 + 5H_2O \text{ or}$$
$$[Cu(H_2O)_4]^{2+} \rightleftharpoons Cu^{2+} + 4H_2O$$

2 Ammonium chloride (Pébal's experiment)

Heat a spatula load of ammonium chloride crystals in a test-tube held at an angle of $45°$. Observe that the crystals sublime and recondense on the upper and cooler parts of the tube.

$$NH_4^+Cl^-(s) \rightleftharpoons NH_3(g) + HCl(g)$$

3 Bismuth chloride oxide (bismuth oxychloride, Gladstone's experiment)

On shaking a few crystals of bismuth(III) chloride with a little distilled water, a colourless solution is obtained. On dilution a white suspension appears. The white suspension dissolves if concentrated hydrochloric (or nitric) acid is added.

$$Bi^{3+}(aq) + H_2O + Cl^- \rightleftharpoons BiOCl(s) + 2H^+$$

4 Iron(III) thiocyanate

To $2\,cm^3$ of iron(III) chloride solution (yellow) add one drop of potassium thiocyanate solution. A red coloration appears. If a few drops of saturated potassium chloride solution are added the red coloration is replaced by the original yellow

$$[Fe(H_2O)_5Cl]^{2+} + SCN^- \rightleftharpoons [Fe(H_2O)_5SCN]^{2+} + Cl^-$$
(etc)

5 Silver and iron ions in solution

To $2\,cm^3$ $0.1\,M$ silver nitrate solution add $4\,cm^3$ of $0.1\,M$ iron(II) sulphate solution. Observe the formation of a precipitate of metallic silver. Pour off $1\,cm^3$ of the solution and to it add one drop of potassium thiocyanate solution: a red coloration appears.

Filter the remainder of the solution and see whether any silver appears in the filtrate due to further reaction. To the filtrate add $2\,cm^3$ of $1\,M$ hydrochloric acid: any white precipitate of silver chloride demonstrates the presence in the filtrate of silver ions (despite the excess of iron(II) ions added above).

To $4\,cm^3$ of silver nitrate solution add a little zinc dust to cause the precipitation of silver. Centrifuge and decant off the supernatant liquid; wash the precipitated silver thoroughly with distilled water. Add $3\,cm^3$ of $0.1\,M$ iron(III) sulphate solution to the silver and shake the mixture well. Then add $2\,cm^3$ of $1\,M$ hydrochloric acid: any white precipitate of silver chloride that forms demonstrates that silver ions have formed from the silver metal.

$$Ag^+ + Fe^{2+} \rightleftharpoons Ag + Fe^{3+}$$

1.53 The Determination of an Equilibrium Constant

$$CH_3COOH + C_2H_5OH \rightleftharpoons CH_3COOC_2H_5 + H_2O$$

1 Various mixtures are left to come to equilibrium at a constant temperature. Dilute hydrochloric acid is a catalyst for the reaction but its presence does not alter the equilibrium constant.

2 Into a thermostatically controlled water-bath, set at 298K (25°C), put a series of corked boiling tubes (150 x 25 mm) containing the following mixtures. A burette or a pipette and a filler should be used to measure these quantities (volumes in cm^3):

Expt	Hydrochloric acid (1M)	Acetic acid (glacial)	Ethanol	Ethyl acetate ethanoate	Water
Density (g/cm^3)	1.00	1.05	0.79	0.92	1.00
Control	1	0	0	0	11
1	1	2	4	2	3
2	1	3	3	3	2
3	1	3	3	2	3
4	1	3	2	3	3
5	1	2	3	3	3
6	1	2	3	4	2

The mixtures must be left for at least three days, preferably seven.

3 After a while pour the contents of the control tube into a conical flask containing about 90 cm^3 of distilled water and add the rinsings; add two drops of phenolphthalein and titrate the solution with 0.2M sodium hydroxide solution. Take a 1 cm^3 sample out of the first boiling tube and put it into a conical flask with about 100 cm^3 of distilled water, add phenolphthalein and titrate as before: repeat the titration with a second sample. Repeat the titrations for the other boiling tubes of the reaction mixture.

4 The titrations in experiments 1–6 are of a mixture of hydrochloric acid and acetic (ethanoic) acid — originally present or formed in the reaction. If these titres are multiplied by 12 and then the titre of the control tube subtracted, the titre for the acetic acid alone is obtained. Calculate the number of millimoles of acetic acid to which this titre corresponds.

Calculate the number of millimoles of acetic acid that were originally present and hence deduce the change in the number of millimoles (gain or loss) of acetic acid when equilibrium is established. Calculate the number of millimoles of ethanol, ethyl acetate (ethanoate) and water originally present: 1M hydrochloric acid is effectively water for the purpose of this section. Using the figure for the change in acetic

acid as equilibrium is established calculate the number of millimoles of ethanol, ethyl acetate and water present at equilibrium. From these values calculate the value of the equilibrium constant, K_C.

$$K_C = \frac{[CH_3COOC_2H_5]\,[H_2O]}{[CH_3COOH]\,[C_2H_5OH]}$$

Introduction to pH and buffer solutions

$$pH = -\log_{10}[H^+]$$

A buffer solution is a solution of known pH which only alters slightly when small quantities of acid or alkali are added to it. To make up a buffer solution place the first named solution in a 250 cm^3 measuring flask and make up to the graduation mark with the second solution. All solutions are 0.1M.

pH	Solution 1		Solution 2
3	4.4 cm^3	CH_3COONa	CH_3COOH
4	38.2 cm^3	CH_3COONa	
5	89.2 cm^3	CH_3COOH	CH_3COONa
6	13.1 cm^3	CH_3COOH	
7	61.0 cm^3	HCl	
8	11.2 cm^3	HCl	
9	1.25 cm^3	HCl	Na_2HPO_4
10	0.9 cm^3	NaOH	
11	8.7 cm^3	NaOH	

1.54 The Determination of pH

1 **Tap-water**
Take $10 \, cm^3$ of tap-water in a test-tube $(100 \times 16 \, mm)$ and add three drops of a universal indicator. Using $10 \, cm^3$ portions of buffer solutions, likewise coloured with indicator, obtain a colour match or an estimate of pH to the nearest integer. Calculate the hydrogen ion concentration in the tap-water.

2 **Eye-wash solutions**
In every laboratory, solutions are kept for washing eyes if there is an accident, e.g. $1\% m/V$ boric acid for alkali in the eye and $1\% m/V$ sodium hydrogencarbonate for acid in the eye. Find the pH of $10 \, cm^3$ portions of these solutions as in (1).

3 **Edible materials**
Study colourless vinegar, a soda mint tablet or some other indigestion tablet dissolved in $10 \, cm^3$ of water, as above.

4 **Carbonic acid**
$$H_2O + CO_2 \rightleftharpoons H_2CO_3 \rightleftharpoons H^+ + HCO_3^- \rightleftharpoons 2H^+ + CO_3^{2-}$$
Saturate $10 \, cm^3$ of water with carbon dioxide: at $273 \, K \, (0°C)$ one volume of carbon dioxide dissolves in one volume of water. Find the pH of the solution as in (1). Acting upon the assumption that carbonic acid is a weak acid and that the only dissociation constant of importance is the first, calculate the value of this dissociation constant, K_a

$$K_a = \frac{[H^+][HCO_3^-]}{[H_2CO_3]} = \frac{[H^+]^2}{[H_2CO_3]}$$

5 **pH and concentration**
Take $10 \, cm^3$ of $1 \, M$ hydrochloric acid and find its pH as in (1). Accurately dilute $10 \, cm^3$ of the solution to $100 \, cm^3$ using pipette and measuring flask. Find the pH of the $0.1 \, M$ solution. Proceed in a like manner to dilute the solution, finding the pH at each stage, until there is no apparent change in the latter.

Plot a graph of pH (vertical axis) against $-\log_{10} c$ of the acid (horizontal axis) and comment upon its shape. Compare the values of pH obtained by using indicators with those obtained by calculation.

Alternatively study sodium hydroxide solution, acetic (ethanoic) acid or ammonia solution.

6 **pH and titration**
By pipette put $25 \, cm^3$ of $0.1 \, M$ sodium hydroxide solution into a conical flask and add three drops of a universal indicator. Into a burette put $0.1 \, M$ hydrochloric acid. Determine the pH of the alkali: if desired transfer $10 \, cm^3$ to a test-tube as in (1) and then return it to the flask. Add $2 \, cm^3$ of acid from the burette and determine the new pH; continue until the pH reaches a constant value.

Plot a graph of pH (vertical axis) against volume of acid added (horizontal axis) and comment upon its shape.

Alternative pairs of solutions are: sodium hydroxide and acetic (ethanoic) acid, ammonia and hydrochloric acid, ammonia and acetic acid, sodium carbonate and hydrochloric acid, and sodium hydroxide (in the burette in this experiment) and phosphoric acid.

1.55 A Study of Buffer Solutions

1 Take about $30 \, cm^3$ of distilled water, boil it and then allow it to cool back to room temperature in a covered beaker. Take $10 \, cm^3$ of this distilled water, add three drops of a universal indicator and then add $0.01 \, M$ hydrochloric acid slowly from a burette until the pH is 6.

2 To $10 \, cm^3$ of a buffer solution of pH = 7 add three drops of a universal indicator and then hydrochloric acid as in (1).

3 Repeat parts (1) and (2) and titrate with $0.01 \, M$ sodium hydroxide solution until the pH is 8. Comment upon the answers obtained in parts (1) to (3).

4 **Acetic acid**
From glacial acetic acid (pure ethanoic acid, density $1.05 \, g/cm^3$) make up $100 \, cm^3$ of $0.001 \, M$ acetic acid. Find its pH to the nearest integer. The Henderson-Hasselbach equation for a buffer solution is

$$pH = \log_{10} \frac{[Salt]}{[Acid]} + pK_a$$

$pK_a \, (-\log_{10} K_a)$ for acetic acid is 4.76 at $298 \, K$ $(25°C)$.

Make up $50 \, cm^3$ of a buffer solution using sodium acetate and glacial acetic acid to the estimated pH (which may or may not be an integer). If a colour match is not obtained make up a second buffer solution . . . it should be possible in this manner to determine the pH to the nearest tenth of a unit.

Compare the value obtained by experiment with that obtained by calculation using Ostwald's Dilution Law:

$$[H^+] = \sqrt{c K_a}$$

where c is the concentration of the solute in mol/dm^3.

II Inorganic Chemistry by Periodic Table Groups

The Periodic Table

Period \ Group	I	II											III	IV	V	VI	VII	0
1																		1 H · 2 He
2	3 Li	4 Be											5 B	6 C	7 N	8 O	9 F	10 Ne
3	11 Na	12 Mg											13 Al	14 Si	15 P	16 S	17 Cl	18 Ar
4	19 K	20 Ca	21 Sc	22 Ti	23 V	24 Cr	25 Mn	26 Fe	27 Co	28 Ni	29 Cu	30 Zn	31 Ga	32 Ge	33 As	34 Se	35 Br	36 Kr
5	37 Rb	38 Sr	39 Y	40 Zr	41 Nb	42 Mo	43 Tc	44 Ru	45 Rh	46 Pd	47 Ag	48 Cd	49 In	50 Sn	51 Sb	52 Te	53 I	54 Xe
6	55 Cs	56 Ba	57 La*	72 Hf	73 Ta	74 W	75 Re	76 Os	77 Ir	78 Pt	79 Au	80 Hg	81 Tl	82 Pb	83 Bi	84 Po	85 At	86 Rn
7	87 Fr	88 Ra	89 Ac**	104 Unq														

	58 Ce	59 Pr	60 Nd	61 Pm	62 Sm	63 Eu	64 Gd	65 Tb	66 Dy	67 Ho	68 Er	69 Tm	70 Yb	71 Lu
*														
**	90 Th	91 Pa	92 U	93 Np	94 Pu	95 Am	96 Cm	97 Bk	98 Cf	99 Es	100 Fm	101 Md	102 No	103 Lr

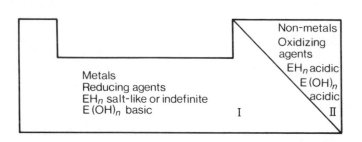

Metals
Reducing agents
EH_n salt-like or indefinite
$E(OH)_n$ basic

I

Non-metals
Oxidizing agents
EH_n acidic
$E(OH)_n$ acidic

II

Binary compounds between elements in set I and in set II: electrovalent
Binary compounds of elements both in set II: covalent
Compounds and mixtures of elements all in set I: alloys

Introduction

1   Counterbalance tubes when putting them in a centri-
    fuge and keep the lid shut whilst the rotor is in
    motion.

2   A 75×10 mm test-tube can contain about 4cm^3; a
    100×16 mm test-tube can contain about 12cm^3 of
    material.

3   The supernatant liquid is the liquid above the precipi-
    tate or crystals.

4   Cut filter-papers for micro-Hirsch funnels with
    scissors, a paper punch or a cork borer. Only use
    small pieces of filter-paper for drying substances.

5   All acids, bases and salts in solution are ionic. When
    writing ionic equations write cations then anions and
    ignore the *structure* of the precipitate unless this
    detail can be checked.

6   The hydration of ions is usually neglected in writing
    equations: it is sometimes hard to decide precisely
    even by X-ray analysis of the solid, and in solution it
    is often indefinite.

7   The quantities taken for experiments should be
    estimated: they are usually in the range 0.001 to
    0.002 of a mole of the material, i.e. about 10^{21} par-
    ticles. A spatula load is taken to mean 0.25–0.5 g.

2.1 The Reactions of Lithium, Sodium and Potassium Metals

Use one small pellet (a 5 mm cube) of freshly-cut lithium (sodium or potassium) for each of the following experiments.

---

1 Place it in a bowl of water behind a safety screen.

---

2 Drop it into a 2 cm³ portion of dry ethanol. Collect a sample of the gas evolved in a test-tube by holding it mouth-to-mouth with the reaction test-tube and then holding it to a flame. What happens when the residue is carefully added to water?

---

3 Place it in a combustion tube, warm gently to melt and then pass oxygen over. Alternatively warm it gently in a deflagrating-spoon and then lower it into a gas-jar of oxygen.

    The oxygen can be prepared by the catalytic decomposition of hydrogen peroxide using manganese(IV) oxide.

---

4 Repeat (3) using dry chlorine, in a fume cupboard. The chlorine can be prepared by dripping concentrated hydrochloric acid on to potassium permanganate crystals.

In each case state the order of reactivity of the three metals. Look up their standard electrode potentials and their first ionization energies. In the light of these values comment upon your results.

2.2 The Reactions of Lithium Ions

To 2 cm³ portions of 1 M lithium chloride solution add the following reagents:
a) Sodium hydroxide solution, until it is present in excess.
b) Dilute ammonia solution, until it is present in excess.
c) Sodium carbonate solution, until it is present in excess. Repeat the addition in the presence of 1 cm³ of dilute ammonia solution.
d) 1 cm³ of dilute hydrochloric acid and then hydrogen sulphide solution.
e) Two spatula loads of ammonium chloride and 1 cm³ of dilute ammonia solution before adding hydrogen sulphide solution or passing in the gas.
f) Disodium hydrogenphosphate solution and, if no effect is observed boil the solution.

    Then perform a flame test: the wavelength of the spectral line responsible for the dominant colour is 671 nm.

Into which group of the qualitative analysis tables would you place lithium?

Precipitation reactions involving sodium and potassium are rarer (see section 3.24). Spectral lines are seen in flame tests at 589 nm for sodium and 404 nm for potassium.

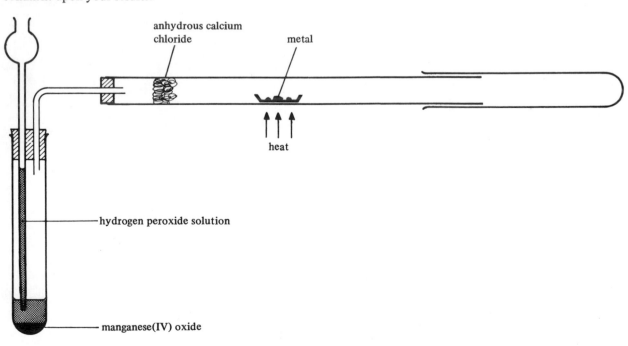

anhydrous calcium chloride

metal

heat

hydrogen peroxide solution

manganese(IV) oxide

Group II

2.3 The Reactions of Magnesium, Calcium and Barium Metals

Use 5 cm of magnesium ribbon (or a granule of calcium; or a small pellet of barium, about a 5 mm cube).

1 Place it in a bowl of water. In the case of magnesium the experiment may be continued in a test-tube containing 2 cm³ of water which is then warmed.

2 Does magnesium burn in steam?

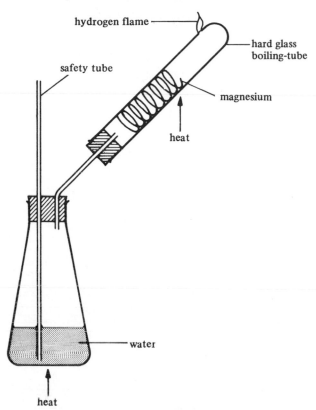

Heat the magnesium with a second burner when the steam is issuing freely from the hole.

3 Put a few drops of mercury(II) chloride on a piece of magnesium ribbon and rub them into the surface using a small piece of paper towel. Discard the paper and thoroughly wash your fingers afterwards. After making this magnesium amalgam does the magnesium now react with water?

4 Does magnesium react with dilute hydrochloric, nitric or sulphuric acids? Check the identity of any gas evolved.

5 Does calcium or barium react with dilute hydrochloric or sulphuric acids? Explain your observations.

6 Does magnesium react at all with sodium chloride or ammonium chloride solutions?

7 Does magnesium react at all with dilute or concentrated sodium hydroxide solution?

In each case state the order of reactivity of these metals. Look up their standard electrode potentials and their first and second ionization energies. In the light of these values comment upon your results.

2.4 The Reactions of Magnesium, Calcium, Strontium and Barium Ions

To 2cm³ portions of 1M solutions of magnesium, calcium, strontium and barium chloride (or nitrate) add the following reagents. Magnesium sulphate may also be used.

a) Sodium hydroxide solution, until it is present in excess.

b) Dilute ammonia solution, until it is present in excess.

c) Ammonium carbonate solution, until it is present in excess.

d) Repeat (c) first having added two spatula loads of ammonium chloride and 2 cm³ of ammonia solution.

e) Two spatula loads of ammonium chloride, 2 cm³ of dilute ammonia solution and then, with stirring, 1 cm³ of disodium hydrogenphosphate solution.

f) To a solution containing magnesium ions add two drops of Magneson I or II and then sufficient dilute sodium hydroxide solution, with stirring, to make the solution alkaline. Magneson I is 4-nitrobenzeneazoresorcinol [4-nitrobenzeneazo-(2,4-dihydroxy)-benzene]

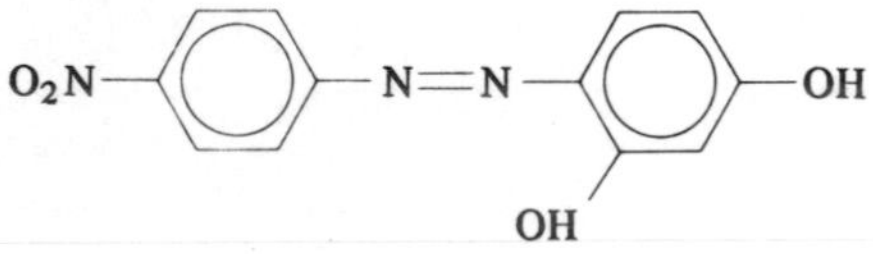

Magneson I

and Magneson II is 4-nitrobenzeneazo-1-naphthol (4-nitrobenzeneazonaphth-1-ol)

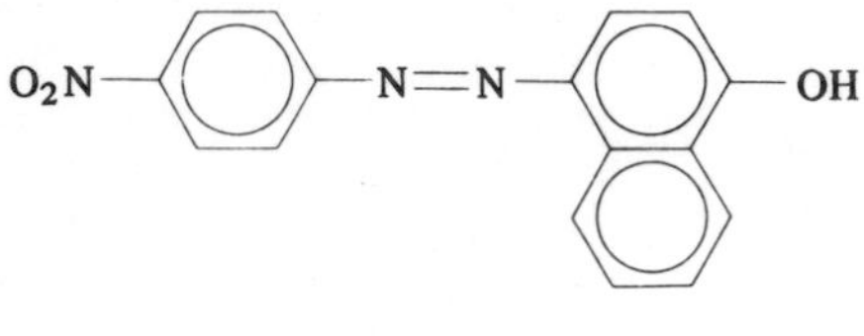

Magneson II

g) Heat one spatula load of a magnesium salt on a charcoal block. Moisten the residue with one drop of cobalt(II) nitrate solution and then reheat it.

h) Dilute sulphuric acid: if there is no visible effect concentrate the solution by boiling it, but do not boil off all the water. To any precipitate add hot concentrated ammonium sulphate solution.

i) Saturated calcium sulphate solution; if there is no visible effect in the case of strontium and barium solutions (only) boil the solution.

j) To 1 cm³ (only) of the solution add 3 cm³ of potassium hexacyanoferrate(II) solution and then three spatula loads of ammonium chloride.

k) Ammonium oxalate (ethanedioate) solution.

l) Potassium chromate(VI) solution.

Then perform a flame test: the wavelengths of the spectral lines responsible for the main colours are at 720 and 555nm for calcium; 674 to 606 and 461nm for strontium and 554 to 514nm for barium.

2.5 The Reactions of Titanium Metal and Titanium(III) Ions

The atoms in a metal are in an oxidation state of zero and hence they may be referred to as, for example, titanium(0).

1 Titanium(0)

Investigate the effect on small pieces of titanium metal of the following reagents.

a) Dilute sulphuric acid, then warm the mixture.

b) Dilute hydrochloric acid, then warm the mixture.

c) Sodium hydroxide solution, then warm the mixture.

2 Titanium(III)

To $2\,cm^3$ portions of a solution containing titanium(III) ions [as produced in part (1)] add the following reagents.

a) Sodium hydroxide solution, until it is present in excess.

b) Dilute ammonia solution, until it is in excess.

c) Hydrogen sulphide (gas or solution).

d) Add hydrogen peroxide dropwise: in the first stage reaction occurs yielding Ti^{4+} in solution; the second stage has not been explained.

e) $1\,cm^3$ of dilute sulphuric acid and, dropwise potassium permanganate [manganate(VII)] solution.

3 Titanium(IV)

To $2\,cm^3$ portions of a solution containing titanium(IV) ions [as produced in part 2 (d)] add the following reagents.

a) Hydrogen peroxide followed by ammonium fluoride solution.

b) Pyrocatechol (benzene-1, 2-diol) solution (1M).

c) Disodium hydrogenphosphate solution.

2.6 The Reactions of Vanadium Metal and Vanadate(V) Ions

1 Vanadium(0)

Investigate the effect on small pieces of vanadium of the following reagents.

a) Dilute sulphuric acid.

b) Hot concentrated sulphuric acid.

c) Cold dilute hydrochloric acid, then warm the mixture.

d) Hot dilute nitric acid.

e) Concentrated sodium hydroxide solution.

2 Vanadium(V)

To $2\,cm^3$ portions of ammonium vanadate(V) add the following reagents.

a) $2\,cm^3$ dilute sulphuric acid and a few drops of hydrogen peroxide.

b) Lead(II) acetate (ethanoate) solution.

c) Barium chloride solution.

d) Copper(II) sulphate solution.

e) Ammonium sulphide solution followed by dilute sulphuric acid.

f) Half a spatula load of ammonium nitrate and $1\,cm^3$ concentrated nitric acid, stir, then add ammonium molybdate [molybdate(VI)] solution.

2.7 The Oxidation States of Vanadium

As you perform the experimental steps complete the statements, state the oxidation numbers (Ox) and describe the colours seen.

1 Dissolve two spatula loads of sodium or ammonium vanadate(V) in about $50\,cm^3$ of distilled water in a conical flask, warming if necessary. To the almost colourless solution add $10\,cm^3$ of dilute sulphuric acid.

$$VO_3^- + H^+ \rightarrow VO_2^+ + H_2O$$

Ox ....

Colour ....

2 Warm the solution to about 340K (67°C) and add a spatula load of sodium metabisulphite [disulphate-(IV)]. Shake the flask.

$$VO_2^+ + SO_3^{2-} + H^+ \rightarrow VO^{2+} + SO_4^{2-} + H_2O$$

Ox ++++   ....

Colour ....

3 Add a further $25\,cm^3$ portion of dilute sulphuric acid and a spatula load of zinc dust. Shake the flask again.

$$VO^{2+} + Zn + H^+ + H_2O \rightarrow [V(H_2O)_6]^{3+} + Zn^{2+}$$

Ox ....

Colour ....

4 Add a second load of zinc dust and lightly cork the flask. Shake the flask yet again.

$$V(H_2O)_6^{3+} + Zn \rightarrow V(H_2O)_6^{2+} + Zn^{2+}$$

Ox ....

Colour ....

2.8 The Reactions of Chromium Metal and Chromium(III) Ions

1 Chromium(0)
Investigate the effect on small pieces of chromium metal of the following reagents.
a) Dilute hydrochloric acid, then warm the mixture.
b) Dilute sulphuric acid, then warm the mixture.
c) Concentrated nitric acid.
d) Sodium hydroxide solution, then warm the mixture.

The presence of air results in solutions containing Cr^{3+} rather than Cr^{2+} ions.

2 Chromium(III)
To $2\,cm^3$ portions of a solution containing chromium(III) ions, e.g. as the chloride, add the following reagents.
a) Sodium hydroxide solution, until it is present in excess.
b) To the solution remaining from test (a) add hydrogen peroxide, boil the mixture and then add acetic (ethanoic) acid and lead(II) acetate (ethanoate) solution.
c) Concentrated ammonia solution dropwise.
d) Ammonium sulphide or hydrogen sulphide solution.
N.B. The precipitate obtained is chromium(III) hydroxide.
e) Sodium carbonate solution. (The precipitate obtained is again the hydroxide).
f) Disodium hydrogenphosphate solution.
g) Two spatula loads of ammonium chloride, and then dilute ammonia solution until the solution is alkaline.

2.9 The Preparation of Chromium(III) Potassium Sulphate-12-Water

This is a double salt otherwise known as chrome alum.

1 Take a spatula load of potassium dichromate [dichromate(VI)] and $3\,cm^3$ of water in a test-tube and warm the tube until the crystals have dissolved. Cool the solution back to room temperature. Next add $0.5\,cm^3$ of concentrated sulphuric acid and again cool the solution back to room temperature.

2 Add $1.5\,cm^3$ of ethanol slowly, in portions, whilst keeping the tube as cold as possible in water. Note the smell of acetaldehyde (ethanal) which has a boiling-point of 294 K (21 °C).

3 Pour the solution obtained in (2) into a small beaker and allow it to crystallize. It is not possible to obtain crystals if the solution is heated above 343 K (70 °C). Filter off the crystals using a Hirsch funnel, rinse them with a little cold water and dry them between filter-papers.

4 Test the crystals or a solution of them for the presence of potassium, chromium(III) and sulphate ions.

2.10 The Dehydration of Chromium(III) Chloride-6-Water by Thionyl Chloride

Many substances containing water of crystallization cannot be dehydrated by heating directly because they undergo hydrolysis, hydrogen chloride being evolved and an oxide or a hydroxide left.

1 Put about 5 g of chromium(III) chloride-6-water in a mortar and use a pestle to reduce it to a fine powder. Place about $15\,cm^3$ of thionyl chloride in the flask fitted with a reflux condenser (figure 4.A) in a fume cupboard and then sprinkle in the powdered chromium chloride. Warm the mixture using a Bunsen burner, tripod and gauze and let it reflux gently for up to two hours. During this time there will be a slow colour change
$$6SOCl_2 + CrCl_3 \cdot 6H_2O \rightarrow CrCl_3 + 6SO_2 \uparrow + 12HCl$$

2 Any excess of thionyl chloride can be decanted away and the crystals rinsed with a little distilled water, in which they are almost insoluble. Alternatively the thionyl chloride can be distilled off (b.p. 342 K, 69 °C), having rearranged the apparatus as in figure 4.C. The solid should be kept in a desiccator which also contains a dehydrating agent, e.g. concentrated sulphuric acid or freshly prepared calcium oxide.

3 Shake a few crystals of the anhydrous product with $2\,cm^3$ of distilled water and then note the effect of adding a few drops of tin(II) chloride solution. If the green crystals in solution are treated with silver nitrate solution quantitative measurements show that the initial precipitate is of only one-third of the chlorine present. The structure of the green crystals has been shown to be $[Cr(H_2O)_4Cl_2]^+Cl^- \cdot 2H_2O$. A violet form which is $[Cr(H_2O)_6]^{3+}(Cl^-)_3$ and a pale green form $[Cr(H_2O)_5Cl]^{2+}(Cl^-)_2 \cdot H_2O$ are also known.

4 To a second portion of the crystals in $2\,cm^3$ of distilled water and in the presence of a drop of tin(II) chloride solution add a spatula load of zinc dust and $1\,cm^3$ of concentrated hydrochloric acid. Centrifuge the test-tube when the reaction has abated and note the colour of the chromium(II) salt produced.

2.11 The Preparation and Properties of Potassium Chromate and Dichromate

1 The preparation of potassium chromate

Take two spatula loads of chromium(III) chloride-6-water and dissolve it in the minimum quantity of warm water. Add sodium hydroxide solution until precipitation ceases and then centrifuge or filter off the precipitate. Heat the precipitate to convert the hydroxide to the oxide. Alternatively start with one spatula load of chromium(III) oxide.

Care: you must wear safety spectacles for the next part of the experiment.

Put the oxide into a crucible and add six pellets of potassium hydroxide and a spatula load of potassium nitrate (this functions as the oxidizing agent). Heat the crucible over a Bunsen burner and, when the potassium hydroxide melts, carefully stir the mixture with a glass rod. Then allow the mixture to cool and when it is quite cold add $6\,cm^3$ of water and warm the crucible gently so that all soluble salts are leached out (extracted). Centrifuge or filter the solution to remove any suspended matter and then boil off half the solvent. Tip the concentrated solution on to a watch-glass and allow it to crystallize. Balance the statements.

$$Cr^{3+} + OH^- \rightarrow Cr(OH)_3 \downarrow$$
$$Cr(OH)_3 \rightarrow Cr_2O_3 + H_2O$$
$$Cr_2O_3 + OH^- + NO_3^- \rightarrow CrO_4^{2-} + NO_2 + H_2O$$

2 The preparation of potassium dichromate

Take three spatula loads of potassium chromate crystals and $10\,cm^3$ of dilute sulphuric acid in a small beaker (or evaporating-basin) and boil off half the water present. Potassium dichromate crystallizes out and it may be purified by recrystallization from hot water.

$$2CrO_4^{2-} + 2H^+ \rightarrow Cr_2O_7^{2-} + H_2O$$

Note that chromium has an oxidation number of six in both chromates and dichromates.

3 The properties of potassium chromate

To $2\,cm^3$ portions of potassium chromate solution add the following reagents.
a) Lead(II) acetate (ethanoate) or nitrate solution.
b) Barium chloride solution.
c) Silver nitrate solution.
d) Dilute sulphuric acid.

4 The properties of potassium dichromate

To $2\,cm^3$ portions of potassium dichromate solution add the following reagents.
a) $1\,cm^3$ dilute sulphuric acid and a spatula load of sodium sulphite [sulphate(IV)] crystals.
b) $1\,cm^3$ dilute sulphuric acid and a few drops of hydrogen sulphide solution (or pass in hydrogen sulphide).
c) $1\,cm^3$ dilute sulphuric acid and a few drops of hydrogen peroxide.
d) $1\,cm^3$ dilute sulphuric acid and $2\,cm^3$ of diethyl ether (ethoxyethane) followed by a few drops of hydrogen peroxide, then stir well for 30 seconds.
Care: ether is flammable.
e) $1\,cm^3$ dilute hydrochloric acid and $1\,cm^3$ of iron(II) chloride solution.
f) $1\,cm^3$ dilute hydrochloric acid and $1\,cm^3$ of tin(II) chloride solution.
g) $1\,cm^3$ dilute sulphuric acid and $1\,cm^3$ potassium iodide solution.
h) Sodium hydroxide solution.

2.12 The Reactions of Manganese Metal and Manganese(II) Ions

1 Manganese(0)
Investigate the effect on small pieces of manganese metal of the following reagents.
a) Dilute hydrochloric acid, then warm the mixture.
b) Dilute sulphuric acid, then warm the mixture.
c) Sodium hydroxide solution, then warm the mixture.

2 Manganese(II)
To $2\,cm^3$ portions of a solution containing manganese(II) ions, e.g. as the sulphate, add the following reagents.
a) Sodium hydroxide solution, until it is present in excess.
b) Dilute ammonia solution, until it is present in excess.
c) $2\,cm^3$ dilute hydrochloric acid, then a few drops of hydrogen sulphide solution (or pass in the gas).
d) Two spatula loads of ammonium chloride, then dilute ammonia until the solution is alkaline and finally a few drops of hydrogen sulphide solution.
e) $1\,cm^3$ concentrated nitric acid and a spatula load of lead(IV) oxide. Boil the mixture then centrifuge. Sodium bismuthate(V) gives the same result in the presence of moderately concentrated nitric acid without boiling.
f) Sodium carbonate solution.

3
Gently heat a spatula load of manganese(II) carbonate in a test-tube and note the ease of decomposition. Sprinkle the residue on a crucible lid and heat it. Suggest reasons for the colour changes observed.

4
Repeat (3) using manganese(II) oxalate (ethanedioate).

N.B. The colours of the carbonate and the oxalate are not typical of manganese(II) salts and the colour of the salts is too pale to be evident in solution.

2.13 The Preparation and Properties of Potassium Permanganate [Manganate (VII)]

1 Preparation
Care: you must wear safety spectacles for part (a) of this preparation.

a) Into a crucible put a spatula load of manganese(IV) oxide, four pellets of potassium hydroxide and half a spatula load of potassium nitrate (employed here as an oxidizing agent). Heat the crucible directly over a Bunsen flame and when the pellets melt carefully stir the molten mass with a glass rod. Then put the crucible down and allow it to cool back to room temperature. Write the equation for the reaction.

b) When the crucible is quite cool add $4\,cm^3$ of water to lixiviate (i.e. to leach out or dissolve) the green compound potassium manganate(VI). Decant the solution into a test-tube and centrifuge it. Put $1\,cm^3$ of the solution on to a watch-glass and after checking that the pH is above seven evaporate the solution to obtain crystals. What are these crystals? Pour the remainder of the solution into a clean test-tube and add dilute sulphuric acid dropwise until the pH is about four. Centrifuge again and decant the solution into a small beaker or on to a watch-glass and allow it to form crystals. (Potassium sulphate stays in solution unless evaporation is prolonged). Filter off the crystals using a Hirsch funnel, wash them with a little cold water and dry them between filter-papers. Write an equation for the oxidation of potassium manganate(VI).

2 Properties
To $1\,cm^3$ portions of potassium permanganate each acidified with $1\,cm^3$ of dilute sulphuric acid add the following reagents. Write equations for reactions.
a) $1\,cm^3$ hydrogen peroxide: test the gas evolved with a glowing splint and to the solution produced add sodium hydroxide solution.
b) $1\,cm^3$ iron(II) sulphate solution followed by $3\,cm^3$ sodium hydroxide solution.
c) $1\,cm^3$ of a solution of oxalic acid (ethanedioic acid) or use sodium oxalate solution. Warm the test-tube and test the gas evolved with calcium hydroxide solution.
d) $1\,cm^3$ sodium sulphite [sulphate(IV)] solution followed by $1\,cm^3$ barium chloride solution and $2\,cm^3$ concentrated hydrochloric acid.
e) A few drops of hydrogen sulphide solution (or pass in the gas).
f) $1\,cm^3$ potassium nitrite [nitrate(III)] solution and then $1\,cm^3$ dilute sulphuric acid and $1\,cm^3$ iron(II) sulphate solution. Cool the test-tube before carefully adding $1\,cm^3$ of concentrated sulphuric acid.
g) $1\,cm^3$ potassium iodide solution then a few drops of starch solution.

3
Perform a flame test on the crystals; to a few of them add a few drops of concentrated hydrochloric acid — observe the evolution of chlorine.

4
To a $1\,cm^3$ portion of potassium permanganate solution, made alkaline with $1\,cm^3$ of sodium carbonate solution, add $1\,cm^3$ of oxalic acid and then warm the mixture. Write an equation for this reaction.

2.14 Manganese in Oxidation States VI, V and III

In this experiment predictions made from reduction potentials can be compared with experimental results. The reduction potential ($E^\ominus$) of a system is an intensive factor and is quoted for solutions that contain $1\,mol/dm^3$ of oxidant and of reductant.

$$
\begin{array}{llll}
MnO_3^- + 2H^+ + e^- & \rightarrow & MnO_2 & + H_2O & 2.5\,V \\
MnO_4^{2-} + 4H^+ + 2e^- & \rightarrow & MnO_2 & + 2H_2O & 2.26\,V \\
MnO_4^{2-} + 2H^+ + e^- & \rightarrow & MnO_3^- & + H_2O & 2.0\,V \\
MnO_4^- + 8H^+ + 5e^- & \rightarrow & Mn^{2+} & + 4H_2O & 1.51\,V \\
Mn^{3+} + e^- & \rightarrow & Mn^{2+} & & 1.51\,V \\
MnO_2 + 4H^+ + e^- & \rightarrow & Mn^{3+} & + 2H_2O & 0.95\,V \\
MnO_3^- + H_2O + e^- & \rightarrow & MnO_2 & + 2OH^- & 0.84\,V \\
MnO_4^{2-} + 2H_2O + 2e^- & \rightarrow & MnO_2 & + 4OH^- & 0.6\,V \\
MnO_4^- + e^- & \rightarrow & MnO_4^{2-} & & 0.56\,V \\
MnO_4^{2-} + H_2O + e^- & \rightarrow & MnO_3^- & + 2OH^- & 0.34\,V \\
MnO_2 + 2H_2O + e^- & \rightarrow & Mn(OH)_3 & + OH^- & 0.2\,V \\
Mn(OH)_3 + e^- & \rightarrow & Mn(OH)_2 & + OH^- & -0.1\,V \\
\end{array}
$$

1 Manganese(VI)

From manganese(IV) oxide and permanganate [manganate(VII)] ions do the reduction potentials predict that manganate [manganate(VI)] ions will be produced in acidic, neutral or alkaline solution?

Take three $4\,cm^3$ portions of potassium permanganate solution: to one add $2\,cm^3$ of dilute sulphuric acid, and to the second add $2\,cm^3$ of sodium hydroxide solution. To each of the three tubes add half a spatula load of manganese(IV) oxide. Shake each tube for a few minutes, centrifuge and inspect. Save the sample containing manganate(VI) ions in solution for the next experiment.

2 Manganese(V)

From manganese(IV) oxide and manganate [manganate(VI)] ions do the potentials predict that manganate(V) ions will be produced in acidic, neutral or alkaline solution?

Take three $1\,cm^3$ portions of the manganate(VI) solution: to one add $1\,cm^3$ of dilute sulphuric acid, and to the second add $1\,cm^3$ of sodium hydroxide solution. To each of the three tubes add half a spatula load of manganese(IV) oxide. Shake each tube for a few minutes, centrifuge and inspect.

Care: you must wear safety spectacles for the next part of this experiment.

Carefully add pellets of potassium (or sodium) hydroxide to $1\,cm^3$ of potassium permanganate solution. There is a colour change and oxygen is evolved. The reducing agent is the hydroxide ion.

$$4OH^- \rightarrow 2H_2O + O_2 + 4e^- \qquad E^\ominus = -0.4\,V$$

Carefully pour away the contents of the tube amid a stream of water and rinse the tube well.

3 Manganese(III)

From manganese(IV) oxide and manganese(II) ions do the redox potentials predict that manganese(III) ions will be produced in acidic, neutral or alkaline solution?

To $2\,cm^3$ manganese(II) sulphate add $2\,cm^3$ of dilute sodium hydroxide solution. Observe and explain what happens.

Put a spatula load of manganese(II) sulphate crystals in $2\,cm^3$ of dilute sulphuric acid and when they have dissolved add $0.5\,cm^3$ of concentrated sulphuric acid. Cool the mixture and then add five drops of potassium permanganate solution. Pour the solution into about $50\,cm^3$ of water and stir well. Observe and explain what happens.

2.15 The Preparation of Ammonium Iron(II) Sulphate-6-Water

This is a double salt, also known as Mohr's salt, $(NH_4)_2 Fe(SO_4)_2 \cdot 6H_2O$.

Take two spatula loads of ammonium sulphate and dissolve it in the minimum quantity of hot water in a test-tube. In a second test-tube boil $5 \, cm^3$ of distilled water to expel any air that has dissolved and then cool the tube under the tap. Acidify the water with five drops of dilute sulphuric acid and use the minimum quantity of this acidified water to dissolve three spatula loads of iron(II) sulphate-7-water in a third test-tube.

Mix the solutions of the two sulphates in a small beaker and allow them to crystallize. Filter off the crystals using a Hirsch funnel, wash them with a little water and dry them between filter-papers.

2.16 The Preparation of Iron(II) Chloride and Iron(III) Chloride

The combustion tube and the boiling tube used as a receiver must be dried thoroughly before proceeding with these experiments which should be done in the fume cupboard.

Iron(II) chloride
Generate a stream of dry hydrogen chloride by dripping concentrated sulphuric acid into concentrated hydrochloric acid. Strong heating of the iron wire (steel wool) is essential as soon as the air has been displaced out of the combustion tube. Not all of the iron(II) chloride produced will volatilize (it sublimes at about 953 K, 680 °C). When the tube is cool the iron and iron(II) chloride remaining can be tapped or scraped out of the tube and examined, e.g. by adding a drop of water.

Iron(III) chloride
Some damp cotton-wool should now be fitted into the top of the first boiling tube. Generate a stream of chlorine by dripping concentrated hydrochloric acid on to crystals of potassium permanganate. The cotton-wool retains any hydrogen chloride that volatilizes but permits the chlorine to proceed to the drying agent. Moderate heating of the iron wire is sufficient and iron(III) chloride vaporizes at about 580K (307°C). To a few of the crystals add a drop of water and note the effect.

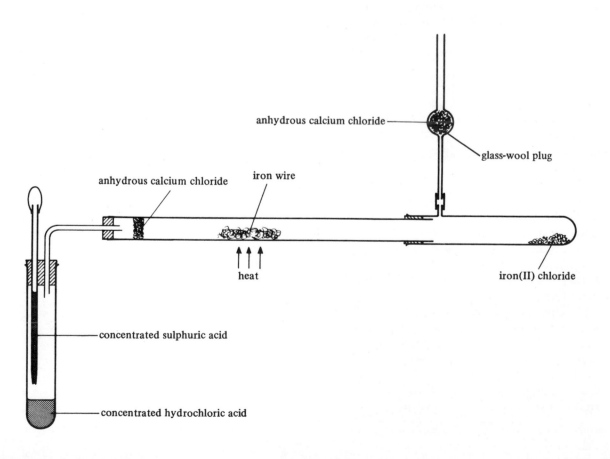

2.17 The Reactions of Iron Metal, Iron(II) Ions and Iron(III) Ions

1 Iron(0)

Investigate the effect on iron filings or small pieces of steel wool of the following reagents.

a) Dilute hydrochloric acid, then warm the mixture.

b) Dilute sulphuric acid, then warm the mixture.

c) Concentrated nitric acid in the cold.

d) Sodium hydroxide solution, then warm the mixture.

e) Air: heat a small ball of steel wool strongly then hold it in tongs in a gas-jar full of air (or oxygen, if available).

f) Sulphur: mix iron filings with sulphur powder and gently warm the mixture.

g) Copper(II) sulphate solution.

2 Iron(II)

To $2cm^3$ portions of a solution containing iron(II) ions, e.g. as the sulphate or double sulphate with ammonium sulphate, add the following reagents.

a) Sodium hydroxide solution, until it is present in excess. Leave to stand and observe later.

b) Dilute ammonia solution, until it is present in excess.

c) Potassium hexacyanoferrate(III) solution $(K^+)_3[Fe(CN)_6]^{3-}$. The crystals of this salt should be rinsed with distilled water and then made up into solution in distilled water which previously has been boiled and allowed to cool. Prussian (or Turnbull's) blue is produced: $(Fe^{3+})_4[Fe^{II}(CN)_6^{4-}]_3 \cdot 14H_2O$

d) Potassium hexacyanoferrate(II) solution $(K^+)_4[Fe(CN)_6]^{4-}$. If the simple salt, iron(II) sulphate is used its crystals should be made into solution as in (c) by stirring the cold solution. The initial white precipitate that is produced is $(K^+)_2[Fe_2^{II}(CN)_6]^{2-}$

e) $1cm^3$ dilute sulphuric acid and potassium permanganate [manganate(VII)] solution dropwise until a faint pink tinge persists. Then add sodium hydroxide solution.

f) A few drops of concentrated nitric acid: warm the mixture and then cool it. Next add some sodium hydroxide solution.

g) $1cm^3$ dilute sulphuric acid and $1cm^3$ hydrogen peroxide solution. Then add sodium hydroxide solution.

h) Fresh colourless ammonium sulphide solution.

i) $1cm^3$ dilute sulphuric acid and $1cm^3$ potassium dichromate [dichromate(VI)] solution.

3

Warm some crystals of iron(II) sulphate in an ignition tube gently at first and then strongly. Observe the colour changes and test the gases evolved: by carefully smelling; with damp litmus paper; with a strongly glowing splint, and by bringing up a glass rod which has been dipped in concentrated ammonia solution.

4 Iron(III)

To $2cm^3$ portions of a solution containing iron(III) ions, e.g. as the chloride, add the following reagents.

a) Sodium hydroxide solution, until it is present in excess.

b) Dilute ammonia solution, until it is present in excess.

c) Potassium (or ammonium) thiocyanate solution. The coloration is produced by the ion $[FeSCN]^{2+}$. Next add a spatula load of sodium fluoride. Explain what you observe.

d) Potassium hexacyanoferrate(III) solution, prepared as in 2(c). A brown coloration due to $Fe^{3+}[Fe^{III}(CN)_6]^{3-}$ is said to be formed but often is (Berlin) green due to contamination with Prussian blue.

e) Potassium hexacyanoferrate(II) solution; Prussian blue is again produced.

f) A spatula load of zinc dust and $1cm^3$ of concentrated hydrochloric acid: allow the reaction to proceed for ten minutes with occasional stirring, then filter and test the solution for iron(II) ions.

g) Pass in hydrogen sulphide, then centrifuge or filter. Examine the residue and the solution.

h) $2cm^3$ tin(II) chloride solution, then sodium hydroxide solution.

i) Fresh colourless ammonium sulphide solution.

j) A spatula load of sodium sulphite [sulphate(IV)]: warm gently and then add sodium hydroxide solution.

2.18 The Reactions of Cobalt Metal and Cobalt(II) Ions

1 Cobalt(0)
Investigate the effect on small pieces of cobalt metal of the following reagents.
a) Dilute hydrochloric acid, then warm the mixture.
b) Dilute sulphuric acid, then warm the mixture.
c) Sodium hydroxide solution, then warm the mixture.

2 Cobalt(II)
To $2\,cm^3$ portions of a solution containing cobalt(II) ions, e.g. as the chloride, add the following reagents.
a) Sodium hydroxide solution, until it is present in excess. Centrifuge, decant off the supernatant liquid and heat the precipitate.
b) Dilute ammonia solution, until it is present in excess.
c) Sodium carbonate solution.
d) $1\,cm^3$ dilute hydrochloric acid, then hydrogen sulphide (gas or solution).
e) Two spatula loads of ammonium chloride and then dilute ammonia solution until the solution is alkaline, and finally hydrogen sulphide (gas or solution).
f) Using only $0.5\,cm^3$ of solution slowly add concentrated hydrochloric acid. The test may be repeated using cobalt(II) nitrate.
g) Heat some of the solution until all the water has been driven off.
h) Two drops of dilute acetic (ethanoic) acid, $2\,cm^3$ potassium nitrite [nitrate(III)] solution and then concentrated potassium chloride solution.
i) Sodium hypochlorite [chlorate(I)] solution dropwise until no further precipitation occurs. The precipitate is cobalt(III) hydroxide. Divide the precipitate into two portions. To the first carefully add concentrated hydrochloric acid to obtain an unstable brown solution of cobalt(III) chloride. Heat the second portion to obtain oxygen and leave a residue of cobalt(II) cobalt(III) oxide.

3
Heat some cobalt(II) carbonate, gently at first then strongly.

4
Heat some borax (disodium tetraborate-10-water, $Na_2B_4O_7 \cdot 10H_2O$) on a piece of platinum wire until a clear glass-like bead is obtained. Then allow the bead to touch a crystal of a cobalt salt and reheat.

2.19 The Preparation and Estimation of Hexaamminecobalt(III) Chloride

This is a complex salt, $[Co^{III}(NH_3)_6]^{3+}(Cl^-)_3$

1 Preparation
a) Into a conical flask put 9g of cobalt(II) chloride crystals and 6g of ammonium chloride and $10\,cm^3$ of distilled water; heat the flask until all the crystals have dissolved. Add four spatula loads of animal charcoal powder: in the absence of charcoal the chloropentaammine would be obtained.
b) The next few steps must be carried out in a fume cupboard. Add $20\,cm^3$ of concentrated ammonia solution and cool the mixture under the tap. Slowly add $20\,cm^3$ of 20-volume hydrogen peroxide. Next heat the mixture to 330–340K (about $60°C$) and maintain at this temperature for about 15 minutes during which time the pink colour should disappear.
c) Cool the mixture under the tap and then in ice. Using a Buchnet funnel and flask filter off the crystals obtained. In a beaker put $3\,cm^3$ of concentrated hydrochloric acid into $80\,cm^3$ of boiling water and then add the crystals. When the crystals have dissolved, leaving the charcoal behind, rapidly filter the hot solution through a fluted filter-paper. To the filtrate add $10\,cm^3$ of concentrated hydrochloric acid and then cool the mixture in ice.
d) Again filter off the crystals using the Buchner funnel. Rinse the crystals with a little acetone (propanone) and dry them by sucking air through for about five minutes. Find the mass of the crystals and hence calculate the percentage yield.
e) The product may be recrystallized as follows: put the crystals in a beaker and add water until in a hot solution all have dissolved. Then, at the boil, add ethanol until there is a faint cloudiness. On cooling slowly yellow crystals are obtained. These crystals can be filtered off as before, washed with two separate portions of acetone and air dried.

2 Estimation of the proportion of Cobalt(III)
$$Co^{3+} + e^- \rightarrow Co^{2+}$$
$$2I^- \rightarrow I_2 + 2e^-$$
$$2S_2O_3^{2-} \rightarrow S_4O_6^{2-} + 2e^-$$
Accurately weigh out about 1.5g of the cobalt salt prepared as above. Put it in a conical flask and add about $20\,cm^3$ of concentrated sodium hydroxide solution. Boil the mixture in a fume cupboard, until the evolution of ammonia has ceased. Cool the resulting suspension of cobalt(III) oxide and then add about 1 g of potassium iodide (or $12\,cm^3$ of 0.5M solution). Acidify the mixture with concentrated hydrochloric acid. Titrate the iodine released with standard (e.g. 0.1M) sodium thiosulphate [thiosulphate(VI)] solution, using starch as an indicator towards the end-point.

2.20 The Reactions of Nickel Metal and Nickel(II) Ions

1 Nickel(0)
Investigate the effect on small pieces of nickel metal of the following reagents.
a) Dilute hydrochloric acid, then warm the mixture.
b) Dilute sulphuric acid, then warm the mixture.
c) Sodium hydroxide solution, then warm the mixture.

2 Nickel(II)
To $2\,cm^3$ portions of a solution containing nickel(II) ions, e.g. as the sulphate, add the following reagents.
a) Sodium hydroxide solution, until it is present in excess. Centrifuge, decant off the supernatant liquid and heat the precipitate.
b) Dilute ammonia solution, until it is present in excess.
c) Sodium carbonate solution.
d) $1\,cm^3$ dilute hydrochloric acid, then hydrogen sulphide (gas or solution).
e) Two spatula loads of ammonium chloride and then dilute ammonia solution until the solution is alkaline, and finally hydrogen sulphide (gas or solution).
f) Using only $0.5\,cm^3$ of solution slowly add concentrated hydrochloric acid.
g) Sodium hypochlorite [chlorate(I)] solution.
h) Two drops of dilute sulphuric acid, four drops of dimethylglyoxime (butanedione dioxime) solution and sufficient dilute ammonia solution to make the mixture alkaline.

3
Heat some nickel(II) carbonate, gently at first then strongly.

4
Heat some borax (disodium tetraborate-10-water) on a piece of platinum wire until a clear glass-like bead is obtained. Then allow the bead to touch a crystal of a nickel salt and reheat.

The penultimate group of the d-block elements are regarded as transition elements.

2.21 The Preparation of Copper(II) Sulphate-5-Water

Warm $4\,cm^3$ of dilute sulphuric acid and then add small portions of copper(II) oxide until no more seems to dissolve. Warm the solution again to check that no more will dissolve.

Centrifuge and transfer the solution to a crucible or a small beaker. Add one drop of dilute sulphuric acid and evaporate the solution down to its crystallization point, which occurs at about half its original volume. Pour the hot solution on to a watch-glass and allow it to crystallize.

2.22 The Reactions of Copper Metal and Copper(II) Ions

1 Copper(0)
Investigate the effect on small pieces of copper metal of the following reagents.
a) Dilute hydrochloric acid, then warm the mixture.
b) Dilute sulphuric acid, then warm the mixture.
c) Concentrated sulphuric acid, then warm the mixture.
d) In a fume cupboard: moderately concentrated nitric acid (1 volume concentrated acid: 1 volume water).
e) In a fume cupboard: concentrated nitric acid.
f) Sodium hydroxide solution, then warm the mixture.
g) Leave a copper turning in $2\,cm^3$ concentrated ammonia solution whilst (2) is studied. Air may be blown through the solution to accelerate any change.

2 Copper(II)
To $2\,cm^3$ portions of a solution containing copper(II) ions, e.g. as the sulphate, add the following reagents.
a) Sodium hydroxide solution, until it is present in excess. Centrifuge, decant off the supernatant liquid and heat the precipitate.
b) Dilute ammonia solution, with stirring, until it is present in excess.
c) Sodium carbonate solution.
d) $1\,cm^3$ dilute hydrochloric acid, then hydrogen sulphide (gas or solution).
e) Using only $0.5\,cm^3$ of solution slowly add concentrated hydrochloric acid. What complex ion was responsible for the original blue colour? What complex ion is now present and what is its colour?

Pour half the solution into a small beaker of water. To the other half add dropwise concentrated ammonia solution. Explain what you observe in each case.
f) Potassium hexacyanoferrate(II) solution.
g) Potassium iodide solution.

3
Heat some copper(II) carbonate, gently at first then strongly.

4
Heat some copper(II) sulphate crystals, gently at first in a test-tube, then strongly on a crucible lid. Explain what you observe.

5
Perform a flame test on a few crystals of a copper salt.

2.23 The Preparation and Properties of Tetraamminecopper(II) Sulphate-1-Water

This is a complex salt, $[Cu(NH_3)_4]^{2+} SO_4^{2-} \cdot H_2O$.

1 Preparation

Dissolve four spatula loads of crystalline copper(II) sulphate in $4 cm^3$ of water, cool the solution and add, with stirring, $2 cm^3$ of concentrated ammonia solution.

Pour the solution into $6 cm^3$ of ethanol in a small beaker and cool the mixture. Filter off the fine crystals in a Hirsch funnel and rinse them with ethanol before drying them between filter-papers.

[The double salt needed for comparison in part (2) can be prepared as in experiment 2.15 substituting copper(II) sulphate-5-water for iron(II) sulphate-7-water.]

2 Properties

The properties of the complex salt $[Cu(NH_3)_4]^{2+} SO_4^{2-} \cdot H_2O$ are to be compared with the properties of the double salt $(NH_4^+)_2 Cu^{2+} (SO_4^{2-})_2 \cdot 6H_2O$ by performing each of these six tests on them.

a) Heat the crystals in a test-tube gently, then moderately strongly, and finally on a crucible lid very strongly. $[Cu(NH_3)_2]^{2+} SO_4^{2-}$ is a pale green salt.
b) To $2 cm^3$ solution add hydrogen sulphide solution (or gas).
c) To $2 cm^3$ solution add potassium hexacyanoferrate(II) solution.
d) Warm $0.5 cm^3$ solution with $3 cm^3$ sodium hydroxide solution.
e) To $2 cm^3$ solution add disodium hydrogenphosphate solution.
f) To one spatula load of the crystals add five spatula loads of anhydrous sodium carbonate and $3 cm^3$ of water. Warm the mixture gently and stir well. Centrifuge and decant off the supernatant liquid. Acidify the solution with dilute hydrochloric acid, warm gently to expel carbon dioxide and then add barium chloride solution.

(At least one test shows a distinct difference between this complex salt and the double salt).

3 Cuprammonium Rayon

Put $2 cm^3$ of copper(II) sulphate solution into a test-tube and add sodium hydroxide solution until no further precipitation takes place. Centrifuge and wash the precipitate three times with distilled water. Dissolve the precipitate in concentrated ammonia solution to obtain Schweitzer's solution: tetraamminecopper(II) hydroxide. Show that the solution will dissolve small shreds of filter- or blotting-paper. Pour the viscous solution produced into moderately concentrated sulphuric acid (made by carefully pouring $20 cm^3$ of concentrated sulphuric acid into $20 cm^3$ water).

2.24 The Preparation and Properties of Copper(I) Oxide

1 Preparation

Heat $2 cm^3$ of copper(II) sulphate solution and $2 cm^3$ of glucose solution with $6 cm^3$ of water in a small beaker until the solution is boiling. Then add, with stirring, $2 cm^3$ of sodium hydroxide solution. Keep the solution hot for another minute.

Centrifuge and wash the precipitate obtained twice with $2 cm^3$ portions of water. Dry the precipitate between filter-papers.

Glucose is an aldehyde $C_5 H_{11} O_5 CHO$ and its oxidation product is gluconic acid $C_5 H_{11} O_5 COOH$. Write an equation for the redox reaction in which copper(I) oxide is prepared.

2 Properties

Copper(I) oxide does not usually yield copper(I) salts by its reactions: often *disproportionation* occurs and the copper(0) formed may react further.
$$2Cu^+ \rightarrow Cu\downarrow + Cu^{2+}$$
To samples of copper(I) oxide add the following reagents.
a) Dilute hydrochloric acid, then warm the mixture.
b) Concentrated hydrochloric acid.
c) Dilute sulphuric acid.
d) Dilute nitric acid.
e) Sodium hydroxide solution.
f) Dilute ammonia solution.
g) Concentrated ammonia solution.
$[Cu(NH_3)_2]^+$ is colourless but is readily oxidized to $[Cu(NH_3)_4(H_2O)_2]^{2+}$ which is blue.

2.25 The Preparation and Properties of Copper(I) Chloride

1 Preparation

Heat one spatula load of copper(II) oxide with $4\,cm^3$ of concentrated hydrochloric acid in a test-tube and when the solution is boiling add a spatula load of copper turnings. Maintain the mixture at the boiling-point for five minutes, adding more concentrated hydrochloric acid periodically to keep the volume constant. Allow the test-tube to cool for a few minutes.

Meanwhile warm $20\,cm^3$ of distilled water to expel dissolved air and then cool it. One crystal of sodium sulphite [sulphate(IV)] may be put into the water because it will undergo preferential oxidation, thus maintaining the copper in its univalent condition. Put $6\,cm^3$ portions of the air-free distilled water into two test-tubes.

Centrifuge the test-tube containing the copper as a dark complex ion and decant the supernatant liquid into the air-free distilled water. Centrifuge and decant off the water. To the precipitates obtained add two $3\,cm^3$ portions of ethanol to remove any water remaining.

Centrifuge and replace the ethanol by two $3\,cm^3$ portions of diethyl ether (ethoxyethane). Shake the precipitates of copper(I) chloride out on to a Hirsch funnel on a Buchner flask and dry by sucking air through with a filter-pump.

Care: diethyl ether is flammable.

2 Properties

a) Is the precipitate soluble in concentrated hydrochloric acid or in saturated sodium chloride solution?
b) Is the precipitate soluble in concentrated ammonia solution? $[Cu(NH_3)_2]^+$ may be formed as a colourless ion but it is rapidly oxidized to the more familiar blue complex ion $[Cu(NH_3)_4(H_2O)_2]^{2+}$
c) Leave a sample out in the air. Oxidation occurs giving basic copper(II) chloride $CuCl_2 \cdot 3Cu(OH)_2$ which is a naturally occurring form of copper (atacamite).
d) Add potassium iodide solution to a sample.
e) Fuse some of the precipitate with an equal quantity of sodium carbonate. Copper(I) carbonate is not stable so copper(I) oxide is produced. Allow the test-tube to cool and then add $3\,cm^3$ of water, stir well and warm gently. Centrifuge and decant off the solution. Acidify the solution with dilute nitric acid, warm to expel carbon dioxide and then add silver nitrate solution.

Copper(I) chloride can be estimated quantitatively by cerium(IV) sulphate (see experiment 6.38).

2.26 The Reactions of Silver(I) Ions

To $2\,cm^3$ portions of silver nitrate solution add the following reagents.

a) Five drops of dilute hydrochloric acid. Then add dilute nitric acid. Centrifuge and to the precipitate add $4\,cm^3$ of dilute ammonia solution. Divide the solution into two portions: to one add dilute nitric acid, to the other add potassium iodide solution.
b) $0.5\,cm^3$ of potassium bromide solution. Then add dilute nitric acid.
c) $0.5\,cm^3$ of potassium iodide solution. Then add dilute nitric acid.
d) $0.5\,cm^3$ dilute sodium hydroxide solution.
e) Dilute ammonia solution dropwise with stirring.
f) Hydrogen sulphide (gas or solution).
g) $0.5\,cm^3$ of potassium chromate solution. Divide the precipitate into two portions: to one add dilute nitric acid, to the other dilute ammonia solution.

2.27 To Determine the Formula of the Amminesilver(I) Ion

Let there be x molecules of ammonia attached as ligands to each silver(I) ion. Two simultaneous equilibria are established.

$$Ag^+ + xNH_3 \rightleftharpoons [Ag(NH_3)_x]^+$$
$$Ag^+ + Br^- \rightleftharpoons AgBr\downarrow$$

Various quantities of 1M ammonia solution are added to $2cm^3$ of 0.1M silver nitrate solution and the total volume made up to $100cm^3$ in each case with distilled water. The amounts of ammonia used are in vast excess over those needed to react with the silver ions and so the amounts of ammonia that form complex ions are negligible. The concentration of complex ions is effectively constant in these experiments because the volume of silver nitrate taken is constant and the stability constant for the ion is high.

The concentration of the uncomplexed silver ion is very small and it is estimated by titration with 0.1M potassium bromide solution. The bromide is run in from the burette until there is a faint permanent cloud of silver bromide; the titre is low so that the total volume at the end is only slightly over $100cm^3$.

The experiment should be done with volumes of ammonia such as 10, 25, 50 and 75 cm^3.

Then
$$K_1 = \frac{[Ag(NH_3)_x^+]}{[Ag^+][NH_3]^x}$$

$$K_2 = [Ag^+][Br^-]$$

from which $\log_{10}[Br] = x \log_{10}[NH_3] + K_3$ where K_1 is the stability constant, K_2 is the solubility product and K_3 is a constant term in the equation derived in accordance with what has been stated above.

Plot a graph of $\log_{10}[Br]$ (vertical axis) against $\log_{10}[NH_3]$ (horizontal axis); the volumes taken can be employed for this purpose. The slope of the graph gives a value for x.

The formula of the ion can also be found by a distribution constant experiment (see 1.21).

The last group of d-block elements are regarded as transition elements.

2.28 The Reactions of Zinc Metal and Zinc(II) Ions

1 Zinc(0)
Investigate the effect on granulated zinc (or pieces of zinc foil) of the following reagents.
a) Dilute hydrochloric acid, then warm the mixture.
b) Dilute sulphuric acid, then warm the mixture.
c) Concentrated sodium hydroxide solution, then carefully warm the mixture.
d) Copper(II) sulphate solution.

2 Zinc(II)
To 2 cm^3 portions of a solution containing zinc(II) ions, e.g. as the sulphate, add the following reagents.
a) Sodium hydroxide solution, until it is present in excess.
b) Dilute ammonia solution, until it is present in excess.
c) 1 cm^3 dilute hydrochloric acid and hydrogen sulphide (gas or solution).
d) Two spatula loads of ammonium chloride, followed by dilute ammonia solution.
e) Two spatula loads of ammonium chloride and 1 cm^3 dilute ammonia solution, followed by hydrogen sulphide (gas or solution).
f) A spatula load of ammonium chloride, followed by disodium hydrogenphosphate solution.
g) Potassium hexacyanoferrate(II) solution.
h) Sodium carbonate solution.

3 Heat zinc carbonate. Identify the gas evolved. Note the colour of the residue when hot and when cold.

2.29 The Reactions of Cadmium Metal and Cadmium(II) Ions

1 Cadmium(0)
Investigate the effect on small pieces of cadmium of the following reagents.
a) Dilute hydrochloric acid, then warm the mixture.
b) Dilute sulphuric acid, then warm the mixture.
c) Sodium hydroxide solution.

2 Cadmium(II)
To 2 cm^3 portions of a solution containing cadmium(II) ions, e.g. as the sulphate, add the following reagents.
a) Sodium hydroxide solution, until it is present in excess.
b) Dilute ammonia solution, until it is present in excess.
c) Repeat (b) with a concentrated solution.
d) 1 cm^3 dilute hydrochloric acid and hydrogen sulphide (gas or solution).
e) Sodium carbonate solution.

3 Heat cadmium carbonate. Identify the gas evolved.

2.30 The Reactions of Mercury(I) Ions and Mercury(II) Ions

1
Carefully use 1 cm^3 portions of solutions of mercury(I) nitrate, $Hg_2(NO_3)_2$, and of mercury(II) chloride, $HgCl_2$, for the following tests. The mercury(I) ion is a complex ion being Hg_2^{2+}. Thoroughly wash your hands and the test-tubes afterwards. The following reagents may be added to each of the mercury solutions.
a) Dilute hydrochloric acid.
b) Hydrogen sulphide (gas or solution).
c) Sodium hydroxide solution, until it is present in excess.
d) Dilute ammonia solution, until it is present in excess.
e) Potassium chromate solution.
f) Tin(II) chloride solution, remembering the result obtained in (a).
g) Potassium nitrite [nitrate(III)] solution.
h) Potassium iodide solution. In the case of mercury(II) chloride stir the suspension and divide it into two portions: centrifuge in order to obtain the precipitate. To one portion of the precipitate add an excess of potassium iodide solution and keep the resultant solution for experiment (2). Put the second portion of the precipitate on a watch-glass over a beaker of boiling water so that the precipitate is dried.

When the precipitate is dry put it into a dry test-tube with a thermometer reading to at least 450 K (about 180°C). Slowly warm the test-tube over a small Bunsen flame and note the approximate transition temperature.

2
To the solution obtained in 1(h) add 1 cm^3 of concentrated sodium hydroxide solution. The solution obtained is Nessler's reagent; it is very sensitive towards ammonia and ammonium salts. Into the solution put one crystal of ammonium chloride. Note the result.

2.31 The Catalytic Effect of d-Block Ions on a Reaction

This experiment is a general approach to a reaction considered elsewhere (1.47).

Into a test-tube put $1\,cm^3$ of potassium iodide solution (0.5M), $1\,cm^3$ of sodium thiosulphate solution [thiosulphate(VI), 0.025M], $1\,cm^3$ of starch solution and $2\,cm^3$ of potassium persulphate solution [peroxodisulphate(VI), 0.02M]. Warm the mixture to about 330 K (about 57 °C). A similar mixture can be left at room temperature and the time noted for the blue colour of starch and iodine to develop.

Into a set of ten test-tubes put further quantities of the mixture and add one drop $(0.05\,cm^3)$ of the following solutions separately to them: chromium(III) chloride, potassium chromate [chromate(VI)], manganese(II) sulphate, potassium permanganate [manganate(VII)], iron(II) sulphate, iron(III) chloride, cobalt(II) chloride, nickel(II) sulphate, copper(II) sulphate and zinc(II) sulphate. The blue colour will appear at different times. Which ions catalyze the reaction? Suggest a mechanism.

2.32 Edta and Other Complex Salts

Edta = ethylenediaminetetraacetic acid, systematically
   bis[di(carboxymethyl)amino] ethane is usually employed as its disodium salt. See the introduction to experiment 6.46.

---

1 Into a set of 15 test-tubes put $2\,cm^3$ samples of the following solutions (approximately 0.1M): magnesium(II) sulphate, calcium(II) chloride, chromium(III) chloride, manganese(II) sulphate, iron(II) sulphate, iron(III) chloride, cobalt(II) chloride, nickel(II) sulphate, copper(II) sulphate, silver(I) nitrate, zinc(II) sulphate, aluminium(III) chloride, tin(II) chloride, tin(IV) chloride and lead(II) nitrate. Add edta solution slowly and from the observations made suggest whether or not any complex between the cation and edta is formed.

---

2 To $2\,cm^3$ of copper(II) sulphate solution add concentrated ammonia solution until the tetraamminecopper(II) ion has been formed, then add edta solution. What do your observations tell you about the relative stabilities of the three complex ions concerned?

---

3 To $2\,cm^3$ of copper(II) sulphate solution add concentrated hydrochloric acid until the tetrachlorocopper(II) ion has been formed, then add edta solution. What do your observations tell you about the relative stabilities of the three complex ions concerned?

---

4 To $2\,cm^3$ copper(II) sulphate solution add sodium salicylate (2-hydroxybenzoate) solution. Is a new complex ion formed?

---

5 To $2\,cm^3$ copper(II) sulphate solution add a mixture of $1\,cm^3$ of pyrocatechol (benzene-1, 2-diol) solution and $1\,cm^3$ of sodium hydroxide solution. Is a new complex ion formed?

---

6 To $1\,cm^3$ copper(II) sulphate solution add successively
   a) concentrated ammonia solution,
   b) sodium salicylate solution,
   c) edta solution,
   d) a mixture of pyrocatechol solution and an equal volume of sodium hydroxide solution.
   What do your observations tell you about the relative stabilities of the complex ions concerned?

2.33 The Preparation of Aluminium Chloride

The apparatus is the same as in experiment 2.16. The experiment must be done in a fume cupboard. Hydrogen chloride or chlorine is generated, dried and then reacts with the hot aluminium turnings to yield aluminium(III) chloride. The apparatus must be dried thoroughly before use.

The aluminium turnings (about 0.5–1 g) should not be heated until it is apparent that the air in the combustion tube has been replaced by the hydrogen chloride. The reaction is exothermic and once started only gentle heating is required. Most of the aluminium chloride sublimes at 456K (183°C) into the receiver and more can be scraped out of the combustion tube. When pure it is white, but most samples of aluminium chloride contain traces of iron so that the product is yellowish.

To a few of the crystals produced add a few drops of water. Note the effect.

2.34 The Preparation of Aluminium Potassium Sulphate-12-Water

This is a double salt $K^+Al^{3+}(SO_4^{2-})_2 \cdot 12H_2O$; it is also known as alum or potash alum. Aluminium sulphate crystals are variously quoted as having 16 or 18 molecules of water of crystallization.

Put two spatula loads of aluminium sulphate crystals and half a spatula load of potassium sulphate with $3\,cm^3$ of water and three drops of dilute sulphuric acid in a crucible or a small beaker and evaporate the solution down to half its original volume. Pour the solution on to a watch glass and allow it to crystallize.

Filter off the crystals using a Hirsch funnel, rinse them with a little water and dry them between filter-papers.

2.35 The Reactions of Aluminium Metal and Aluminium(III) Ions

1 **Aluminium(0)**
Investigate the effect on aluminium turnings (or small pieces of foil) of the following reagents.
a) Dilute hydrochloric acid, then warm the mixture.
b) Dilute sulphuric acid, then warm the mixture.
c) Concentrated nitric acid in the cold.
d) Concentrated sodium hydroxide solution, then carefully warm the mixture.
e) Put ten drops of mercury(II) chloride solution on a piece of aluminium sheet (or foil). Using a small piece of paper towel or emery paper rub the solution into the metal surface then discard the paper and wash your hands thoroughly. Allow the piece of foil to stand in air; observe and possibly touch it when a reaction is seen to start.

2 **Aluminium(III)**
To $2\,cm^3$ portions of a solution containing aluminium(III) ions, e.g. as the sulphate or chloride, add the following reagents.
a) Sodium hydroxide solution, until it is present in excess.
b) Dilute ammonia solution, until it is present in excess.
c) Sodium carbonate solution; the precipitate obtained is aluminium hydroxide. Explain why.
d) Hydrogen sulphide (gas or solution), the precipitate, if any is obtained, is aluminium hydroxide.
e) One drop of litmus and then dilute ammonia solution. On centrifuging a 'lake' is obtained: a precipitate with a dye adsorbed on its surface.
f) Repeat (e) using alizarin.

2.36 The Properties of Carbonates

The carbonates of ammonium and the Group I metals are soluble in water. The carbonates of sodium and potassium are the only ones which are very stable to heat. The carbonates of aluminium, iron(III), etc. are not precipitated in aqueous solution by sodium carbonate (or hydrogencarbonate) solution; instead the hydroxide is formed.

1 Tests for carbonates

a) Dilute nitric acid

Dilute nitric, hydrochloric or sulphuric acid may be used for this test.

All carbonates when treated with a dilute mineral acid yield carbon dioxide. The acid employed must be stronger than carbonic acid, i.e. have a greater value for its ionization constant. Complications occur firstly, if the metal chloride or sulphate is insoluble in water (the reaction period then may be too short to achieve positive identity), and secondly, with some natural carbonates which require powdering and the use of warm acid.

b) Barium chloride solution

To a solution of the suspected carbonate (if it is soluble) add barium chloride solution, or alternatively add calcium chloride solution.

c) Silver nitrate solution

To a solution of the suspected carbonate (if it is soluble) add silver nitrate solution. Then boil the suspension.

d) Colour

Examine a selection of carbonates. They sometimes have a different colour from the other salts of the metal.

e) The action of heat

Examine by heating a selection of metallic carbonates testing with calcium hydroxide solution for any carbon dioxide evolved. On a small scale the gas above the heated carbonate can be sucked into a previously deflated teat pipette and then the gas expelled through the calcium hydroxide solution. Alternatively, a drop of calcium hydroxide solution can be held on a glass rod immediately above the heated carbonate. Note any colour changes.

f) Precipitation reactions

To a selection of metallic salts in solution, e.g. as in experiment 2.32, add sodium carbonate solution. These may give basic carbonates as precipitates but to the eye no difference is apparent from the normal carbonate.

2 Tests for hydrogencarbonates

The hydrogencarbonates of sodium and potassium are known in the solid state. The hydrogencarbonates of the Group II metals and iron(II) exist in aqueous solution. All hydrogencarbonates are unstable to heat.

a) Magnesium sulphate

To a cold solution add magnesium sulphate solution. Carbonates give immediate precipitates; hydrogencarbonates do not give precipitates until, on warming, the magnesium hydrogencarbonate formed has decomposed to give magnesium carbonate.

b) The action of heat

Solutions of hydrogencarbonates on heating evolve carbon dioxide.

Sodium and potassium hydrogencarbonate evolve carbon dioxide on heating and the solid residue evolves carbon dioxide on treatment with dilute nitric (or hydrochloric or sulphuric) acid.

2.37 The Preparation and Properties of Carbon Monoxide

1 Preparation

The experiment must be conducted in a fume cupboard. Concentrated sulphuric acid is allowed to drip on to sodium formate (methanoate) and the mixture is heated gently. After collecting three or four test-tubes of the gas pass the gas into a test-tube containing hot copper(II) oxide — see 3(d).

2 Physical properties

a) Look up or state the density of carbon monoxide relative to that of air.

b) Would it be safe to try to collect carbon monoxide by displacement of air?

c) What is the solubility of carbon monoxide in water? Invert a test-tube full of the gas open end downwards into a beaker of water.

3 Chemical properties

a) Pour $1\,cm^3$ of calcium hydroxide solution into a test-tube full of the gas. Shake it: is there any noticeable effect?

b) Apply a burning splint to a test-tube full of the gas. After the flame has gone out pour in $1\,cm^3$ of calcium hydroxide solution and shake the test-tube.

c) Test the gas with damp litmus paper.

d) Pass the gas over hot copper(II) oxide. If there is sufficient of an excess of carbon monoxide, the gases issuing from the hole can be ignited after all the air has been displaced from the test-tube.

2.38 The Preparation of a Silica Garden

This experiment is used to demonstrate osmosis.

A solution containing 5–25% of sodium silicate in water is prepared and put in a beaker. Into it are dropped large crystals (a powder is no use) of salts of various elements, especially d-block elements. The crystals fall to the bottom of the beaker and then appear to grow upwards during the next few days. Suitable substances are: chromium(III) sulphate-6-water, manganese(II) sulphate-4-water, iron(III) chloride-6-water, cobalt(II) chloride-6-water, nickel(II) sulphate-7-water, copper(II) sulphate-5-water, aluminium(III) sulphate-18-water and aluminium(III) potassium sulphate-12-water.

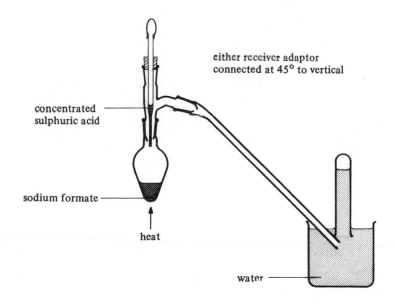

concentrated sulphuric acid

sodium formate

heat

water

either receiver adaptor connected at 45° to vertical

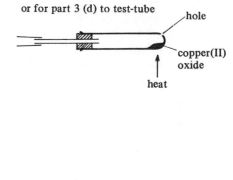

or for part 3 (d) to test-tube

hole

copper(II) oxide

heat

2.39 The Preparation of Tin(IV) Chloride

Set up the apparatus as shown. Chlorine is generated by dripping concentrated hydrochloric acid on to potassium permanganate crystals. When the air has been displaced from the flask the tin (three pieces of granulated) is warmed. Because of moisture in the apparatus the crystals of tin(IV) chloride will be hydrated ($5H_2O$).

By suitable modifications anhydrous tin(IV) chloride can be prepared.

Care: use a fume cupboard.

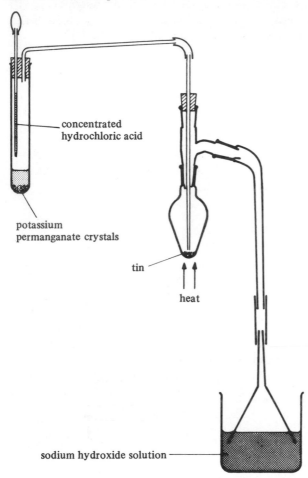

concentrated hydrochloric acid

potassium permanganate crystals

tin

heat

sodium hydroxide solution

2.40 The Preparation of Tin(IV) Iodide

Put two pieces of granulated tin in a boiling-tube and add about $5 \, cm^3$ of a solution of iodine in carbon tetrachloride (tetrachloromethane). Shake the tube occasionally and when the solution has become colourless pour it on to a watch-glass in a fume cupboard. The carbon tetrachloride evaporates away leaving bright orange crystals.

2.41 The Reactions of Tin Metal, Tin(II) Ions and Tin(IV) Ions

1 **Tin(0)**
Investigate the effect on small pieces of granulated tin (remember that tin is expensive) of the following reagents.
a) Dilute hydrochloric acid, then warm the mixture.
b) Dilute sulphuric acid, then warm the mixture.
c) Dilute nitric acid.
d) In a fume cupboard: concentrated nitric acid.
e) Concentrated sodium hydroxide solution, then carefully warm the mixture.

2 **Tin(II)**
To $2 \, cm^3$ portions of a solution containing tin(II) ions, e.g. as the chloride, add the following reagents. Tin(II) salt solutions are usually acidified to minimize hydrolysis.
a) Sodium hydroxide solution, until it is present in excess.
b) Ammonia solution, until it is present in excess.
c) Hydrogen sulphide (gas or solution). Shake the suspension and divide it into three test-tubes: to one add concentrated hydrochloric acid, to the second add lithium hydroxide solution and to the third ammonium sulphide solution.
d) $1 \, cm^3$ mercury(II) chloride solution.
e) A piece of zinc (granulated or foil).
f) $1 \, cm^3$ potassium dichromate [dichromate(VI)] solution mixed with $1 \, cm^3$ dilute sulphuric acid.
g) $1 \, cm^3$ potassium permanganate [manganate(VII)] solution mixed with $1 \, cm^3$ dilute sulphuric acid.

3 **Tin(IV)**
To $2 \, cm^3$ portions of a solution containing tin(IV) ions, e.g. as the chloride, add the following reagents. Tin(IV) salt solutions are usually acidified to minimize hydrolysis.
Tests **(a) – (d)** are as for tin(II) above.
e) A piece of zinc. After a few minutes decant off the solution and add $1 \, cm^3$ mercury(II) chloride solution to it: explain what you observe.

2.42 The Simultaneous Preparation of Lead(II) Nitrate and Lead(IV) Oxide

Put 1 cm³ of water and 1 cm³ of concentrated nitric acid into a test-tube and then sprinkle in a spatula load of trilead tetraoxide [dilead(II) lead(IV) oxide]. Warm the mixture and stir it with a glass rod. The red solid should react completely and be replaced by a brown suspension. If some red solid remains add some more moderately concentrated nitric acid.

Centrifuge and pour the supernatant liquid on to a watch-glass or small beaker to allow the lead(II) nitrate to crystallize.

Wash the precipitate of lead(IV) oxide twice with water and then dry it between filter-papers.

Filter off the crystals of lead(II) nitrate using a Hirsch funnel, rinse them with a little water and dry them between filter-papers.

2.43 The Preparation of Lead(II) Iodide

To 2 cm³ of lead(II) acetate (ethanoate) solution add 2 cm³ of potassium iodide solution. Shake the mixture, warm it then cool it and centrifuge. Discard the clear solution and wash the precipitate twice with 2 cm³ portions of cold distilled water.

Add 3 cm³ of water, shake and heat the mixture to boiling: note that the solution is colourless despite the intense colour of the precipitate. Quickly centrifuge and transfer the solution to a clean tube and cool it under the tap. Shake and pour the suspension on to a small filter-paper. Allow the filter-paper to dry on a watch-glass on a beaker full of boiling water.

2.44 The Reactions of Lead Metal and Lead(II) Ions

1 **Lead(0)**
Investigate the effect on small pieces of lead foil of the following reagents.
a) Dilute hydrochloric acid, then warm the mixture.
b) Dilute sulphuric acid, then warm the mixture.
c) Dilute sodium hydroxide solution, then warm the mixture.
d) The atmosphere: warm some lead in a crucible.
Care: use a fume cupboard.

2 **Lead(II)**
To 2 cm³ portions of a solution containing lead(II) ions, e.g. as the nitrate or the acetate (ethanoate) — the only two salts that are appreciably soluble — add the following reagents.
a) Sodium hydroxide solution, until it is present in excess.
b) Ammonia solution, until it is present in excess.
c) Hydrogen sulphide (gas or solution).
d) Dilute hydrochloric acid, warm the mixture and then allow it to cool.
e) Concentrated hydrochloric acid, until it is present in excess.
f) Potassium iodide solution, until it is present in excess.
g) Dilute sulphuric acid.
h) Potassium chromate solution.
i) Sodium carbonate solution.
j) The brown ring test for a nitrate: this is complicated by the fact that iron(II) sulphate with a lead(II) salt gives a white precipitate.

Therefore to the lead(II) nitrate or acetate solution add an excess of sodium carbonate solution. Centrifuge and decant the supernatant liquid into a clean tube. To the solution add dilute sulphuric acid until there is no more effervescence and after stirring the mixture is acidic. Warm the solution to expel all carbon dioxide and then cool it **thoroughly**. Next add a fresh solution of iron(II) sulphate and, holding the tube at 45°, add a little concentrated sulphuric acid.

3 Investigate the action of heat on lead(II) oxide, dilead(II) lead(IV) oxide and lead(IV) oxide: heat gently at first, then strongly.

4 Investigate the action of heat on lead(II) carbonate and lead(II) nitrate.

5 Perform a flame test on a few crystals of a lead salt.

2.45 The Preparation and Properties of Ammonia

1 Preparation

Set up the apparatus shown in a fume cupboard. Into the reaction vessel put a slurry of roughly equal masses of ammonium chloride and calcium hydroxide. Heat the mixture and collect several test-tubes full of the gas. The solution in water may also be prepared as suggested above.

2 Physical properties

a) Look up or state the density of ammonia relative to that of air.

b) Why cannot concentrated sulphuric acid, phosphorus(V) oxide or calcium chloride be used to dry the gas?

c) What is the solubility of ammonia in water? Why is it dangerous to try to collect ammonia by a delivery tube straight into the water? Invert a test-tube full of the gas open end downwards into a beaker of water.

3 Chemical properties of the gas

a) Shake a test-tube containing 1 cm³ of concentrated hydrochloric acid and then quickly drain off the liquid. Hold the test-tube containing fumes of hydrogen chloride mouth-to-mouth with one of the tubes containing ammonia.

b) Hold damp neutral litmus paper in the gas (or test the solution).

c) Hold damp turmeric paper in the gas.

d) Hold filter-paper dampened with mercury(I) nitrate in the gas.

e) Hold filter-paper dampened with manganese(II) sulphate and hydrogen peroxide in the gas.

4 Chemical properties of the solution

a) Nessler's reagent: see experiment 2.30(2).

b) Add ammonia solution until it is present in excess to solutions of the following metallic salts (approximately 0.1M): sodium chloride, potassium chloride, magnesium sulphate, calcium hydrogencarbonate, calcium chloride, chromium(III) chloride, manganese(II) sulphate, iron(II) sulphate, iron(III) chloride, cobalt(II) chloride, nickel(II) sulphate, copper(II) sulphate, silver(I) nitrate, zinc(II) sulphate, cadmium(II) sulphate, mercury(I) nitrate, mercury(II) chloride, aluminium(III) chloride, tin(II) chloride, tin(IV) chloride, lead(II) nitrate, antimony(III) chloride and bismuth(III) chloride.

Compare the results you obtain with those obtained by the reaction of sodium hydroxide solution on these metallic salts.

2.46 The Properties of Ammonium Salts

1

Warming an ammonium salt (either the solid or a solution) with a strong base (sodium, potassium or calcium hydroxide) releases ammonia.

Why is calcium hydroxide usually employed for this purpose in the solid state?

2

Nessler's reagent: see experiment 2.30(2).

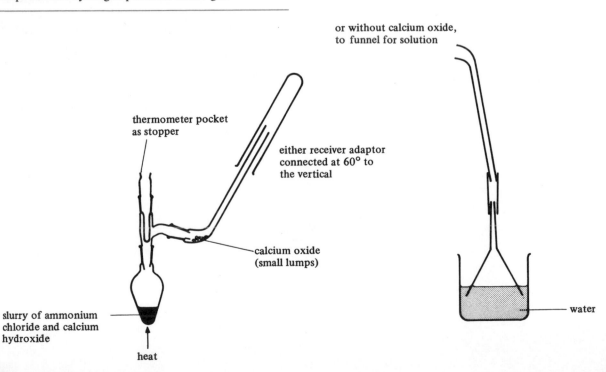

or without calcium oxide, to funnel for solution

thermometer pocket as stopper

either receiver adaptor connected at 60° to the vertical

calcium oxide (small lumps)

slurry of ammonium chloride and calcium hydroxide

heat

water

2.47 The Reactions of Nitrite [Nitrate(III)] Ions

To $2 \, cm^3$ portions of a solution containing nitrite ions e.g. as the sodium salt, add the following reagents.
a) Dilute hydrochloric acid. Identify the gases evolved.
b) A mixture of dilute sulphuric acid and iron(II) sulphate solution.
c) Potassium iodide solution, followed by dilute sulphuric acid.
d) A mixture of dilute sulphuric acid and potassium permanganate solution.
e) Ammonium chloride, then boil the mixture.
f) Amidosulphuric (usually called sulphamic) acid. In this reaction all nitrite ions are converted into nitrogen and no traces of nitrate ions are produced. Thus this is a good method of recognizing a nitrate in the presence of a nitrite.
g) See experiment 2.48(2b).

2.48 The Reactions of Nitrate [Nitrate(V)] Ions

1 On the solid
Three tests can be carried out on a solid nitrate, e.g. sodium nitrate.
a) The action of heat
Exception: ammonium nitrate is too dangerous to heat alone to completion.
  Three equations summarize the reactions that may be observed.

$$2E^I NO_3 \rightarrow 2E^I NO_2 + O_2 \qquad (E^I = Na, K, etc.)$$
$$2E^{II}(NO_3)_2 \rightarrow 2E^{II}O + 4NO_2 + O_2 \ (E^{II} = Cu, Pb, Fe, etc.)$$
$$2E^I NO_3 \rightarrow 2E^I + 2NO_2 + O_2 \quad (E^I = Ag, Hg)$$

b) Concentrated sulphuric acid
To a spatula load of the nitrate add $1 \, cm^3$ of concentrated sulphuric acid and then carefully heat the mixture. Hydrogen nitrate is evolved usually accompanied by nitrogen dioxide.
c) Concentrated sulphuric acid and copper
Proceed as in **(b)** and then when heating slide a copper turning down the side of the test-tube. The hydrogen nitrate reacts with the copper turning and intensifies the evolution of nitrogen dioxide. The solution becomes blue because of the presence of copper(II) nitrate.

2 Solution tests
To $2 \, cm^3$ portions of a solution containing nitrate ions, e.g. as the sodium salt, add the following reagents.
a) The brown ring test
Acidify the solution with dilute sulphuric acid and then add a fresh solution of iron(II) sulphate. Hold the test-tube at an angle of $45°$ and carefully pour concentrated sulphuric acid down the side so that it forms a lower layer.
  If the metal sulphate is insoluble in water the experiment must be modified — see experiment 2.44(2j)
b) Devarda's alloy
Add a spatula load of Devarda's alloy (copper 50%, aluminium 45%, zinc 5%) or zinc dust or aluminium powder on their own. Then add $2 \, cm^3$ of sodium hydroxide solution and warm the solution. Reduction of a nitrate [and a nitrite (nitrate(III))] occurs yielding ammonia.

2.49 The Preparation of Phosphorus Trichloride

The apparatus must be thoroughly dried. Set up the apparatus as shown in a fume cupboard. Pass carbon dioxide or natural gas through the apparatus to displace the air inside it. Using a knife cut several small pieces of white phosphorus under water in an evaporating-basin: use tongs to hold the phosphorus. Shake the phosphorus dry and put it into the apparatus.

Pass carbon dioxide or natural gas through the apparatus again to displace any air that has gained entry and to carry out any water vapour. Then pass dry chlorine over the phosphorus which may be gently warmed.

The phosphorus trichloride (b.p. 349 K, 76°C) distils into the cooled receiver. The phosphorus trichloride can be purified by redistillation using the apparatus as above less the chlorine generator and drier, but with the addition of a thermometer.

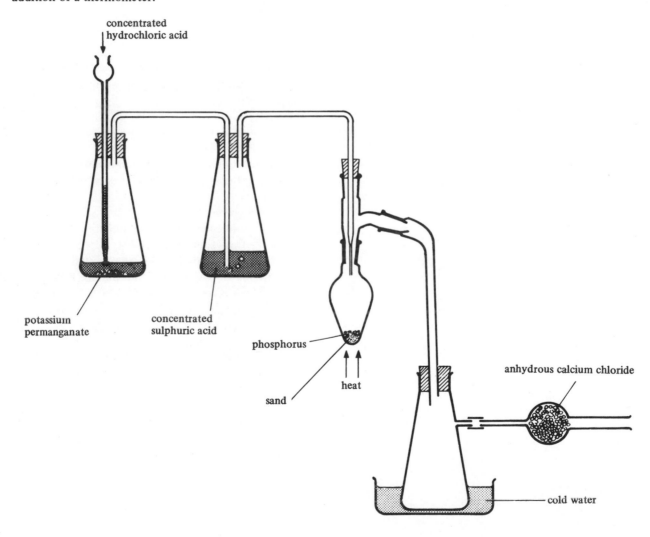

concentrated hydrochloric acid

potassium permanganate

concentrated sulphuric acid

phosphorus

sand

heat

anhydrous calcium chloride

cold water

2.50 The Preparation of Phosphorus Pentachloride

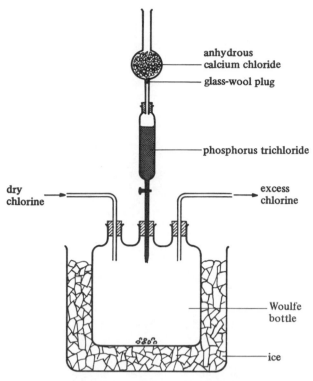

- anhydrous calcium chloride
- glass-wool plug
- phosphorus trichloride
- dry chlorine
- excess chlorine
- Woulfe bottle
- ice

The apparatus must be thoroughly dried. Set up the apparatus as shown in a fume cupboard. Allow a slow stream of dry chlorine to pass through the apparatus as phosphorus trichloride is slowly dripped in.

2.51 The Preparation and Properties of the Phosphates of Sodium

1 **Sodium dihydrogenphosphate-2-water (the orthophosphate) $NaH_2PO_4 \cdot 2H_2O$**
Put $5 cm^3$ of 1M phosphoric acid in a small beaker and add two drops of methyl orange. From a burette add 1M sodium hydroxide solution until, on stirring, the indicator has just changed its colour.

Evaporate the solution down to half its volume and leave it to crystallize on a watch-glass resting on a beaker full of boiling water. Filter off the crystals using a Hirsch funnel, rinse them with a little water and dry them between filter-papers.

2 **Disodium hydrogenphosphate-12-water. $Na_2HPO_4 \cdot 12H_2O$**
To $5 cm^3$ of 1M phosphoric acid add double the quantity of 1M sodium hydroxide solution found in (1). Obtain the crystals as before.

3 **Trisodium phosphate-12-water, $Na_3PO_4 \cdot 12H_2O$**
To $5 cm^3$ of 1M phosphoric acid add three times the quantity of 1M sodium hydroxide solution found in (1). Obtain the crystals as before.

4 **Sodium metaphosphate [sodium polytrioxophosphate(V)], $NaPO_3$**
Take some sodium dihydrogenphosphate-2-water (prepared as above) in a test-tube and heat it very gently until a glass-like melt remains.

5 **Sodium pyrophosphate [sodium heptaoxodiphosphate(V)] $Na_4P_2O_7$**
Take some disodium hydrogenphosphate-12-water prepared as above in a test-tube and heat it very gently until no more water vapour is evolved and a fine powder remains.

6 **Properties of the phosphates**
To $2 cm^3$ portions of solutions of each of these phosphates in water add the following reagents.
a) $1 cm^3$ of concentrated nitric acid and $1 cm^3$ of ammonium molybdate [molybdate(VI)] solution. Stir them and allow them to stand. If no precipitate has formed after ten minutes put the tubes in a beaker of warm water (at 310K, 37°C).
b) Silver nitrate solution.
c) Copper(II) sulphate solution.
d) Barium chloride solution.
e) A spatula load of ammonium chloride, $2 cm^3$ of dilute ammonia solution and $2 cm^3$ of a solution containing magnesium ions.
f) Repeat test (a) in the presence of $1 cm^3$ of tin(II) chloride solution.
g) Iron(III) chloride solution.
h) Zirconyl nitrate [zirconium(IV) nitrate oxide, $ZrO(NO_3)_2$] solution.

2.52 The Reactions of Antimony(III) and Bismuth(III) Ions

To 2 cm^3 portions of antimony(III) chloride and bismuth(III) nitrate solutions (it may be necessary to include a small quantity of the corresponding acid) add the following reagents.

1 **To both solutions**
  a) Hydrogen sulphide (gas or solution). Shake the suspension and divide it between two test-tubes. To one add concentrated hydrochloric acid, to the other add lithium (or sodium) hydroxide solution.
  b) Water in excess, then add a concentrated solution of the corresponding acid.
  c) Sodium hydroxide solution, until it is present in excess.
  d) Ammonia solution, until it is present in excess.
  e) Potassium iodide solution, until it is present in excess.
  f) Sodium carbonate solution.

2 **To the antimony(III) solution only**
  a) Granulated zinc or iron wire or granulated tin.
  b) 1 cm^3 concentrated nitric acid and 1 cm^3 ammonium molybdate [molybdate(VI)] solution, then warm gently.
  c) A modification of Marsh's test for arsenic. In a fume cupboard fit a conical flask with a stopper carrying an absorption tube full of anhydrous calcium chloride leading to a jet. Suspend into the conical flask a small test-tube containing a spatula load of an antimony compound in solution. In the flask generate hydrogen from granulated zinc and moderately concentrated hydrochloric acid.

  Collect ignition tubes full of gas generated in the flask and when the gas burns smoothly upon ignition all the air has been displaced from the apparatus. The gas issuing from the jet can now be ignited safely. On holding a clean crucible or evaporating-basin over the flame only the condensation of steam is observed.

  Tip the flask so that the antimony-containing solution mixes with the zinc and acid. The colour of the flame now changes and a metallic deposit is obtained on the crucible. The deposit is not soluble in sodium hypochlorite [chlorate(I)] solution.

  Under similar conditions substances containing arsenic give a metallic deposit which is soluble in sodium hypochlorite solution.

2.53 The Preparation and Properties of Hydrogen Peroxide

1 Preparation

Make a suspension of four spatula loads of barium peroxide-8-water in $5\,cm^3$ of water and then add it to $5\,cm^3$ of dilute sulphuric acid until the mixture is only faintly acidic on stirring. The mixture should be cooled, preferably with ice and water. Centrifuge and use the supernatant liquid for the tests below.

---

2 Properties

Use $2\,cm^3$ portions of hydrogen peroxide solution for each of the following tests.

a) Investigate the influence of the following on the rate of decomposition: manganese(IV) oxide, sodium hydroxide solution, dilute sulphuric acid and silver oxide. A glowing splint should be used to test for oxygen evolved.

b) Add $1\,cm^3$ of dilute hydrochloric acid and $1\,cm^3$ of sodium sulphite [sulphate(IV)] solution. Then add barium chloride solution followed by some concentrated hydrochloric acid.

c) Add $1\,cm^3$ of dilute sulphuric acid and $1\,cm^3$ of potassium iodide solution. A few drops of starch solution may be added to confirm your first observation.

d) Add $1\,cm^3$ dilute sulphuric acid and $1\,cm^3$ of a fresh solution of iron(II) sulphate. Then add sodium hydroxide solution.

e) Add $1\,cm^3$ of sodium hypochlorite [chlorate(I)] solution and identify the gas evolved.

f) Add $1\,cm^3$ of dilute sulphuric acid and $1\,cm^3$ of potassium permanganate [manganate(VII)]. Identify the gas evolved. To the final solution add sodium hydroxide solution until a precipitate (or merely an opalescence) is observed.

g) Add $1\,cm^3$ of dilute sulphuric acid and $1\,cm^3$ of potassium dichromate [dichromate(VI)] solution. A reaction occurs which has two stages. The situation can be clarified by extinguishing all flames nearby and adding $1\,cm^3$ of diethyl ether (ethoxyethane) before the potassium dichromate solution.

2.54 The Reactions of Sulphide, Sulphite [Sulphate(IV)] and Sulphate [Sulphate(VI)] Ions

The sodium salts can be used in aqueous solution. To $2\,cm^3$ portions of each of the solutions add the following reagents.

a) Dilute hydrochloric acid, then warm the mixture and identify the pungent smelling gases evolved in two cases.

N.B. If sulphur dioxide is bubbled into calcium hydroxide solution a precipitate appears but later disappears if passage of the gas is prolonged. Confusion with carbon dioxide, however, should not arise. Why?

b) Barium chloride solution, followed by concentrated hydrochloric acid.

c) Silver nitrate solution.

d) Lead(II) nitrate [or acetate (ethanoate)] solution.

e) A few drops of a solution of iodine (and potassium iodide) in water.

f) $1\,cm^3$ of dilute sulphuric acid and $1\,cm^3$ of potassium permanganate [manganate(VII)] solution.

g) $1\,cm^3$ of dilute sulphuric acid and $1\,cm^3$ of potassium dichromate [dichromate(VI)] solution.

h) To sodium sulphide solution (only) add a few drops of sodium hydroxide solution and then sodium nitroprusside [pentacyanonitrosylferrate(III)] solution, $(Na^+)_2\,[Fe(CN)_5\,NO]^{2-}$.

2.55 The Preparation and Properties of Sodium Thiosulphate-5-Water

This compound is also known as thio and as hypo, $Na_2S_2O_3 \cdot 5H_2O$.

1 Preparation

Put $4\,cm^3$ of water in a small beaker and add two spatula loads of sodium sulphite-7-water crystals and one spatula load of crushed roll sulphur. Heat the mixture to its boiling-point and maintain it at that temperature for 20 minutes. Stir the solution occasionally and replace any water lost by evaporation.

Allow the solution to cool for a few minutes and then centrifuge it in a test-tube to obtain the clear solution of sodium thiosulphate. Put the clear solution in the beaker again and boil off at least half the water present. Pour the hot solution on to a watch-glass and allow it to crystallize; it may require 'seeding', i.e. the addition of a crystal of sodium thiosulphate from the stock bottle.

Filter off the crystals using a Hirsch funnel, rinse them with a little water and dry them between filter-papers.

2 Properties

a) The crystals readily yield a supersaturated solution in water. They are so soluble that on heating they will melt and dissolve in their own water of crystallization.

Put crystals to a depth of 2–3 cm in a clean test-tube and warm them gently. When a solution has been obtained cork the test-tube and allow it to cool without disturbance. It may be cooled under the tap if the tube is held still. Seed the solution by dropping in a crystal and observe what happens. Feel the test-tube when you observe an occurrence within it.

b) To two spatula loads of the crystals add $2\,cm^3$ of dilute hydrochloric acid. Observe what happens and then warm the tube. Identify the gas evolved.

c) To $2\,cm^3$ of a solution of sodium thiosulphate add a few drops of a solution of iodine (and potassium iodide) in water.

d) To $2\,cm^3$ of sodium thiosulphate solution add barium chloride solution.

e) To $2\,cm^3$ of silver nitrate solution add sodium thiosulphate solution until it is present in excess.

f) Make small quantities of suspensions of silver chloride, bromide and iodide by adding silver nitrate solution to the appropriate halides in solution. Add an excess of sodium thiosulphate solution to each of these suspensions and observe what happens.

2.56 The Reactions of Fluoride Ions

1 In a fume cupboard: to one spatula load of sodium or calcium fluoride in a test-tube add $2\,cm^3$ of concentrated sulphuric acid. Warm the mixture gently. The hydrogen fluoride formed fumes in moist air and the test-tube assumes a greasy appearance because its surface is etched by hydrogen fluoride.

   A moist glass rod held in the vapours will also show signs of being etched.

2 To $2\,cm^3$ of sodium fluoride solution add $1\,cm^3$ of silver nitrate solution.

3 To $2\,cm^3$ of sodium fluoride solution add $1\,cm^3$ of calcium chloride solution.

4 [Reminder: an iron(III) salt with potassium thiocyanate gives a red coloration.]

   To $2\,cm^3$ of a concentrated solution of sodium fluoride in water add a few drops of iron(III) chloride solution followed by a few drops of potassium thiocyanate solution.

   Repeat the above test but add a few drops of concentrated hydrochloric acid before the potassium thiocyanate solution.

2.57 The Preparation and Properties of Hydrogen Bromide

1 **Preparation**
   Set up the apparatus illustrated below in a fume cupboard. Collect four test-tubes full of the gas and then make some solution.

2 **Physical properties**
   a) Look up or state the density of hydrogen bromide relative to that of air.
   b) What is the solubility of hydrogen bromide in water? Would it be safe to try to collect hydrogen bromide by a delivery tube straight into the water? Invert a test-tube full of the gas open end downwards into a beaker of water.

3 **Chemical properties of the gas**
   a) Shake a test-tube containing $1\,cm^3$ of concentrated ammonia solution and then quickly drain off the liquid. Hold the test-tube containing fumes of ammonia mouth-to-mouth with one of the tubes containing hydrogen bromide.
   b) Hold damp neutral litmus paper in the gas (or test the solution).
   c) Into a test-tube full of the gas put a burning splint.

4 **Chemical properties of the solution**
   a) To $2\,cm^3$ of the solution add $1\,cm^3$ of silver nitrate solution. Divide the suspension between two test-tubes: to one add dilute nitric acid, to the other add concentrated ammonia solution.
   b) To $2\,cm^3$ of the solution add two spatula loads of manganese(IV) oxide. Warm the mixture in a fume cupboard. Observe the gas evolved: test it with damp neutral litmus paper.
   c) To $2\,cm^3$ of the solution add two spatula loads of sodium carbonate-10-water. Identify the gas evolved.
   d) To $2\,cm^3$ of the solution add $1\,cm^3$ of sodium hypochlorite [chlorate(I)] solution.

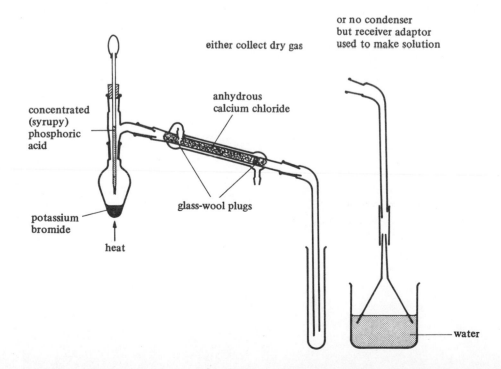

either collect dry gas

or no condenser but receiver adaptor used to make solution

concentrated (syrupy) phosphoric acid

anhydrous calcium chloride

glass-wool plugs

potassium bromide

heat

water

2.58 The Reactions of Chloride, Bromide and Iodide Ions

1 Tests on the solids

To a spatula load of sodium chloride, potassium bromide and potassium iodide separately add each of the following reagents.

a) $2\,cm^3$ concentrated sulphuric acid, then gently warm the mixture. Test the gas evolved with damp neutral litmus paper, with a glass rod moistened with a concentrated solution of ammonia and by cautiously smelling it.

b) In a fume cupboard add two spatula loads of manganese(IV) oxide and $2\,cm^3$ of concentrated sulphuric acid. Stir the mixture and then gently warm it. Observe the colour of the gas evolved, then test it with damp neutral litmus paper and with a piece of filter-paper soaked in fluorescein solution.

---

2 Tests on the solutions

To $2\,cm^3$ portions of solutions of sodium chloride, potassium bromide and potassium iodide individually add each of the following reagents.

a) Silver nitrate solution. Observe the colour of the suspension and then divide it between two test-tubes. To one add dilute nitric acid, to the other add concentrated ammonia solution.

b) Lead(II) nitrate [or acetate (ethanoate)] solution.

c) Chlorine water or a mixture of $1\,cm^3$ of sodium hypochlorite [chlorate(I)] solution and $1\,cm^3$ dilute hydrochloric acid.

d) Bromine water.

e) Sodium nitrite [nitrate(III)] solution.

f) Copper(II) sulphate solution.

g) Mercury(II) chloride solution.

h) Hydrogen peroxide and dilute sulphuric acid. Observe the effect, if any, and then add a few drops of starch solution.

III Inorganic Chemistry by Qualitative Analysis

Introduction

1 The general procedure is based on two assumptions:
a) there is only one cation and one anion in the simple salt provided for analysis,
b) no phosphate separation is necessary: if the substance is a phosphate then the cation is in Groups I or II, or is found by a flame test.

2 A precipitate should be washed before proceeding.

3 The results should be written up as you proceed, negative results as well as positive ones.

4 If you are in difficulties think, then ask if necessary.

5 It is sometimes useful to perform:
a) a blank test with water,
b) a control test with the substance your unknown is suspected to be.

6 Counterbalance tubes when putting them in a centrifuge and keep the lid shut whilst the rotor is in motion.

7 These analysis tables may be used for small scale ('semi-micro') analysis if the quantities stated are halved.

8 A 75×10 mm test-tube can contain about $4 \, cm^3$; a 100×16 mm test-tube can contain about $12 \, cm^3$. A spatula load is taken to mean 0.25–0.5 g of material.

9 Information written within brackets $\{ \ \}$ is not usually required for 'A' level but may be of interest to students afterwards.

10 Be careful not to confuse analytical groups with periodic table groups.

The Principles

These tables are for the most part based on the solubilities of substances and their colour reactions.

The solubility product of the salt, say A^+B^-, is $[A^+][B^-]$ where [ ] are used to denote the concentration of the ion in mol/dm^3. This value must be exceeded by the product of the concentrations of the ions for precipitation to take place. Very insoluble substances have low solubility products. The solubility product of a substance is a constant at a given temperature.

Precipitation of a substance is often avoided by diminishing the concentration of a component ion, e.g. the selective precipitation of sulphides in Groups II and IV.

Dissolution of a substance is obtained in the following ways.
a) By diminishing the concentration of an ion of the substance in the solution above the precipitated substance, e.g.

$$AgCl \, (s) \rightleftharpoons Ag^+ \, (aq) + Cl^- \, (aq)$$

Silver ions are removed by adding ammonia solution.

$$Ag^+ + 2NH_3 \rightarrow [Ag(NH_3)_2]^+$$

b) By heating, e.g. the separation of lead(II) chloride from silver chloride and mercury(I) chloride in the examination of Group I.
c) By conversion of the relevant compound into a soluble substance, e.g. in Group IIA adding moderately concentrated nitric acid to lead(II) sulphide.

$$3PbS + 8H^+ + 2NO_3^- \rightarrow 3S\downarrow + 3Pb^{2+} + 4H_2O + 2NO\uparrow$$

Writing-up Analysis

The examiner assesses your practical examination according to the quality and quantity of the information you give him. This is not an excuse for untidy or sloppy work at the bench because that is likely to lead to errors. The writing-up can be spread over four half-pages.

Date

Substance 1

Test	Result	Inference
Preliminary Tests		
1 Appearance		
2 Solubility		
3 Heat		
4 Flame test		
5 Sodium hydroxide solution		
Anion Tests		
1 Dilute hydrochloric acid		
2 Concentrated sulphuric acid		
3 Boil with excess sodium carbonate solution, test resultant solution: (a) Dilute nitric acid and silver nitrate solution . . .		
Cation Separation		
1 Dilute hydrochloric acid		
2 Hydrogen sulphide into acidic solution . . .		
Cation Identification		
1 Group examination:		
2 Confirmatory tests:		

Conclusion
Substance 1 is . . .

3.1 Appearance

Quick observation may give some indication as to the identity of the substance.

Colour	Inference
White, even when moistened	Absence of all substances below
Blue or violet	Cu^{2+}, Cr^{3+}
Green	Fe^{2+}, Cu^{2+}, Ni^{2+}, Cr^{3+}
Yellow	Fe^{3+}, PbO
Red or pink or crimson	Mn^{2+}, Fe_2O_3, HgO, Pb_3O_4, Co^{2+}
Black	CuO, MnO_2, Fe_3O_4, Co_3O_4
Brown	Ag_2O, PbO_2, $MnCO_3$, CdO

The possibilities are not all listed, e.g. the sulphides and hydroxides that are mentioned later have not been included. The colour may only appear on moistening: the hydration of ions is usually disregarded when writing equations.

$[Cu(H_2O)_4]^{2+}$ blue
$[Cr(H_2O)_6]^{3+}$ violet

Chromium and copper are also green in some basic salts.

$[Fe(H_2O)_6]^{2+}$ pale green
$[Ni(H_2O)_6]^{2+}$ deep green
$[Fe(H_2O)_6]^{3+}$ yellow
$[Mn(H_2O)_4]^{2+}$ pink
$[Co(H_2O)_6]^{2+}$ pink

The colour of a substance is due to the interaction of light with the electrons in the various energy levels of the orbitals.

3.2 Solubility

Find the first possible solvent for the substance trying these in order:
a) cold, then hot water,
b) cold, then hot dilute hydrochloric acid,
c) cold, then hot concentrated hydrochloric acid.
Use two spatula loads of substance and up to $6\,cm^3$ of solvent. Keep the solution for Section 3.9.

If the substance is soluble in water test the solution with pH paper to see if any hydrolysis has taken place: this indicates the strength of the acid or of the base.

If the substance is insoluble in all of the above solvents presume that it is a Group I substance. If a hot solvent or concentrated hydrochloric acid has been used cool the solution before presuming the absence of Group I.

If acid has to be used to obtain a solution a gas may be evolved and the tests in section 3.6 can be carried out now.

Difficulty may be experienced with substances that readily hydrolyze, e.g. salts of tin(IV), antimony(III) and bismuth(III).

3.3 The Action of Heat

Heat two spatula loads of the substance in a dry ignition tube (50 × 10mm, soft glass). This is not a conclusive test and so it must be done quickly; have damp pH paper, calcium hydroxide solution, a splint, your nose and eyes ready!

Result	Inference
Sublimation (fairly common)	e.g. NH_4^+ salts, Hg compounds
Decrepitation (common for crystals)	e.g. $Pb(NO_3)_2$, NaCl
Carbon dioxide evolved (calcium hydroxide solution test)	CO_3^{2-}, HCO_3^-
Carbon monoxide evolved, possibly with carbon dioxide (combustion test)	$C_2O_4^{2-}$
Ammonia evolved (smell, alkaline)	NH_4^+
Oxygen evolved (splint test)	e.g. NO_3^-
Oxygen and nitrogen dioxide evolved (brown, acidic, smell, splint test)	NO_3^-
Sulphur dioxide evolved (smell, acidic)	SO_3^{2-}, SO_4^{2-}
Hydrogen chloride evolved (smell, acidic, fumes with ammonia)	Hydrated Cl^-
Charring	$C_2O_4^{2-}$, CH_3COO^-
Hydrogen sulphide evolved (smell, lead acetate test)	S^{2-}, $S_2O_3^{2-}$
Water evolved, often accompanied by a permanent colour change:	Hydrate of:
Blue → white → black	Cu^{2+}
Green → yellow → red (black)	Fe^{2+}
Green → yellow → green	Ni^{2+}
Yellow → red (black)	Fe^{3+}
Violet → green	Cr^{3+}
Pink → blue → black	Co^{2+}
Temporary colour change:	
White ⇌ yellow	Zn^{2+}, Sn^{2+}, Na_2SO_3
Yellow ⇌ red	Pb^{2+}, Bi^{3+}
Red ⇌ black	Hg^{2+}, Fe^{3+}
Red ⇌ yellow	HgI_2

Potassium and sodium are the only elements to form solid hydrogencarbonates. Carbonates, except those of sodium and potassium when pure, usually decompose yielding the oxide by releasing carbon dioxide.

$$\text{e.g. } M^{II}CO_3 \rightarrow M^{II}O + CO_2$$

Carbon dioxide tested using calcium hydroxide solution.

$$Ca^{2+} + 2OH^- + CO_2 \rightarrow CaCO_3\downarrow + H_2O$$

Oxalates (ethanedioates) usually decompose to give the carbonate or oxide.

$$C_2O_4^{2-} \rightarrow O^{2-} + CO\uparrow + CO_2\uparrow$$

Ammonium salts usually undergo thermal dissociation, so that if the test is prolonged both alkaline and acidic gases may be detected.

$$\text{e.g. } NH_4Cl \rightleftharpoons NH_3\uparrow + HCl\uparrow$$

Nitrates [nitrates(V)] usually undergo thermal decomposition to give the oxide except those of sodium and potassium which on gentle heating give the nitrite [nitrate(III)].

$$2NO_3^- \rightarrow 2NO_2^- + O_2\uparrow \ (K^+, Na^+)$$
$$4NO_3^- \rightarrow 2O^{2-} + 4NO_2\uparrow + O_2\uparrow \ (\text{most})$$
$$2M^+ + 2NO_3^- \rightarrow 2M + 2NO_2\uparrow + O_2\uparrow \ (Ag^+, Hg^{2+}, Hg_2^{2+})$$

and dangerously

$$NH_4NO_3 \rightarrow N_2O\uparrow + 2H_2O\uparrow$$

Sulphates on heating may decompose to the oxide and sulphur trioxide, the latter perhaps undergoing thermal dissociation.

$$\text{e.g. } 2FeSO_4 \cdot 7H_2O \rightarrow Fe_2O_3 + SO_2\uparrow + SO_3\uparrow + 14H_2O\uparrow$$

Hydrated chlorides may hydrolyze on heating, e.g. $AlCl_3 \cdot 6H_2O$.

$$Cl^- + H_2O \rightarrow OH^- + HCl\uparrow$$

Organic compounds on strong heating may char, yielding carbon, etc.

Most hydrates lose all their water of crystallization on heating. Permanent colour changes may result from dehydration of a salt and decomposition to an oxide.

$[Cu(H_2O)_4]^{2+}$ blue → Cu^{2+} white → CuO black
$[Fe(H_2O)_6]^{2+}$ green → Fe^{2+} yellow → Fe_2O_3 red (black)
$[Ni(H_2O)_6]^{2+}$ green → Ni^{2+} yellow → NiO dark green
$[Fe(H_2O)_6]^{3+}$ yellow → Fe_2O_3 red (black)
$[Cr(H_2O)_6]^{3+}$ violet → Cr_2O_3 green
$[Co(H_2O)_6]^{2+}$ pink → Co^{2+} blue → Co_3O_4 black

Temporary colour changes may result from changes in crystal structure, e.g. zinc oxide, etc.

3.4 The Flame Test

Pour $2\,cm^3$ of concentrated hydrochloric acid into a crucible and use this acid to clean the nichrome wire, then heat the substance on the wire moistened with clean acid. This is not a conclusive test and should be done rapidly.

Result	Inference
Brilliant yellow (persistent, masked by cobalt glass)	Na^+
Yellow (sparks)	Fe compound, NH_4^+
Lilac (crimson through cobalt glass)	K^+
Brick red or dull orange (usually with sodium contamination, light green through cobalt glass)	Ca^{2+}
Light green (persistent)	Ba^{2+}
Green with blue centre	Cu^{2+}
White flash	Pb or Sn compound, Sb^{3+}, $\{As^{3+}\}$
$\{$ Crimson (almost unchanged through cobalt glass)	Sr^{2+}, Li^+ $\}$

Except for copper and iron compounds the compounds containing these ions are not usually coloured.

The chlorides of the metals are usually their most volatile salts and so give the best results. The colours are due to energy changes of the electrons.

3.5 The Action of Sodium Hydroxide Solution

This is a vital test because the presence or absence of an ammonium radical must be established at this stage: presence means that the cation tests described later can be ignored.

Warm two spatula loads of the substance with $4\,cm^3$ of dilute sodium hydroxide solution. The evolution of ammonia (smell, alkaline to litmus, forms white clouds with hydrogen chloride) proves the presence of an ammonium salt.

$$NH_4^+ + OH^- \rightarrow H_2O + NH_3\uparrow$$
$$NH_3(g) + HCl(g) \rightarrow NH_4Cl(s)$$

The effect on the substance in other ways, e.g. colour, solubility, is often useful subsidiary evidence. When all, if any, of the ammonia has been evolved add a small quantity of Devarda's alloy or zinc dust or aluminium powder and check that the solution is alkaline: hydrogen will be produced on warming but if ammonia is also evolved a nitrite [nitrate(III)] or nitrate [nitrate(V)] is present.

Side effects with sodium hydroxide, and (possibly) carbonate, include:
$Hg\downarrow + HgO\downarrow$, $Ag_2O\downarrow$ black
$HgO\downarrow$ yellow
$Mg(OH)_2\downarrow$, $Bi(OH)_3\downarrow$, $\{Cd(OH)_2\downarrow\}$ white
$Ni(OH)_2\downarrow$ green
$Cu(OH)_2\downarrow$ blue, on warming $\rightarrow CuO\downarrow$ black
$Cr(OH)_3\downarrow$ mauve or green, with excess alkali $\rightarrow [Cr(OH)_6]^{3-}$ pale yellow solution
$Fe(OH)_2\downarrow$ white (occasionally seen) $\rightarrow Fe_3(OH)_8\downarrow$ green, in air $\rightarrow Fe(OH)_3\downarrow$ red-brown
$Mn(OH)_2\downarrow$ white, in air $\rightarrow MnO_2\cdot H_2O\downarrow$ brown
$Co(OH)_2\downarrow$ blue, on warming $\rightarrow$ pink, in air $\rightarrow$ black

$Sn^{2+}, Sn^{4+}, Sb^{3+}, Al^{3+}, Zn^{2+}, Pb^{2+}, \{As^{3+}\}$ may initially give white precipitates which are soluble in excess alkali as complex anions: $HSnO_2^-$, SnO_3^{2-}, SbO_2^-, $[Al(OH)_6]^{3-}$, $[Zn(OH)_4]^{2-}$, $[Pb(OH)_4]^{2-}$, $\{AsO_2^-, AsO_3^{3-}\}$

The use of Devarda's alloy (Cu 50%, Al 45%, Zn 5%) is often a more reliable test than the brown ring test.
$$8Al + 3NO_3^- + 21OH^- + 18H_2O \rightarrow 8[Al(OH)_6]^{3-} + 3NH_3\uparrow$$
$$2Al + NO_2^- + 5OH^- + 5H_2O \rightarrow 2[Al(OH)_6]^{3-} + NH_3\uparrow$$

3.6 The Action of Dilute Hydrochloric Acid

To $4 \, cm^3$ of dilute hydrochloric acid add two spatula loads of the solid; warm if no reaction in the cold.

	Result	Inference	
a)	Carbon dioxide evolved (slightly acidic, no smell, calcium hydroxide solution turns milky)	CO_3^{2-}	$2H^+ + CO_3^{2-} \rightarrow H_2O + CO_2\uparrow$ $Ca^{2+} + 2OH^- + CO_2 \rightarrow CaCO_3\downarrow + H_2O$
b)	Oxides of nitrogen evolved (brown fumes, acidic, smell)	NO_2^-	$2H^+ + 2NO_2^- \rightarrow H_2O + NO\uparrow + NO_2\uparrow$
c)	Hydrogen sulphide evolved (smell, slightly acidic, lead acetate (ethanoate) solution turns black)	S^{2-}	$2H^+ + S^{2-} \rightarrow H_2S\uparrow$ $Pb^{2+} + S^{2-} \rightarrow PbS\downarrow$
d)	Sulphur dioxide evolved on warming (potassium chromate [chromate(VI)] or dichromate [dichromate(VI)] and dilute sulphuric acid turns green, potassium permanganate [manganate(VII)] and dilute sulphuric acid turns colourless)	SO_3^{2-}	$2H^+ + SO_3^{2-} \rightarrow H_2O + SO_2\uparrow$ $2CrO_4^{2-} + 2H^+ \rightarrow Cr_2O_7^{2-} + H_2O$ $Cr_2O_7^{2-} + 3SO_3^{2-} + 8H^+ \rightarrow 2Cr^{3+} + 3SO_4^{2-} + 4H_2O$ $2MnO_4^- + 5SO_3^{2-} + 6H^+ \rightarrow 2Mn^{2+} + 5SO_4^{2-} + 3H_2O$ (Caution: hydrogen sulphide also reduces potassium dichromate and permanganate; sulphur is an additional product but may not be seen on a filter-paper.)
e)	Sulphur deposited in cold (white), sulphur dioxide evolved on warming	$S_2O_3^{2-}$	$2H^+ + S_2O_3^{2-} \rightarrow S\downarrow + H_2O + SO_2\uparrow$
f)	Acetic (ethanoic) acid evolved on warming (vinegary smell, acidic). If in doubt check by using dilute sulphuric acid or by warming with ethanol and a drop of concentrated sulphuric acid giving ethyl acetate, recognized by its odour.	CH_3COO^-	$H^+ + CH_3COO^- \rightarrow CH_3COOH\uparrow$ $C_2H_5OH + CH_3COOH \rightleftharpoons CH_3COOC_2H_5 + H_2O$
g)	Chlorine evolved on warming (smell, bleaches litmus or pH paper, yellow-green)	Oxidizing agent	$2Cl^- \rightarrow Cl_2\uparrow + 2e^-$

3.7 The Action of Concentrated Sulphuric Acid

This test should be omitted if the previous test produced positive results and a simple salt only is being analyzed; it must be done for mixtures. It must **not** be done at home on an unknown substance.

To two spatula loads of the solid add 2 cm³ of concentrated sulphuric acid; warm carefully if no reaction occurs in the cold.

(All results obtained in section 3.6 will be obtained again here if a mistake has been made).

	Result		*Inference*
a)	Hydrogen chloride evolved (acidic, smell, fumes in air, white clouds with ammonia)	Cl^-	$\underset{\text{halide}}{H_2SO_4 + X^-} \rightarrow HSO_4^- + HX\uparrow$
b)	Hydrogen bromide, mixed with bromine and sulphur dioxide, evolved (as above but brown fumes)	Br^-	If hydrogen bromide also: $H_2SO_4 + 2HBr \rightarrow 2H_2O + Br_2\uparrow + SO_2\uparrow$
c)	Hydrogen iodide, mixed with iodine, sulphur dioxide and hydrogen sulphide, evolved (as above but violet fumes)	I^-	If hydrogen iodide also: $H_2SO_4 + 2HI \rightarrow 2H_2O + I_2 + SO_2\uparrow$ $H_2SO_4 + 6HI \rightarrow S\downarrow + 3I_2 + 4H_2O$ $H_2SO_4 + 8HI \rightarrow 4H_2O + 4I_2 + H_2S\uparrow$ The iodine may be evolved as a gas and/or remain in suspension.
d)	Hydrogen nitrate [nitrate(V)] and nitrogen dioxide evolved (acidic, smell, oily drops condense at top of tube, brown fumes—intensified by adding a copper turning). Distinction from bromine obtained by testing with fluorescein on filter-paper: bromine with fluorescein goes to eosin (pink).	NO_2^- and NO_3^-	$H_2SO_4 + 2NO_2^- \rightarrow SO_4^{2-} + H_2O + NO\uparrow + NO_2\uparrow$ $H_2SO_4 + NO_3^- \rightarrow HSO_4^- + HNO_3\uparrow$ $4HNO_3 \rightarrow 2H_2O + 4NO_2\uparrow + O_2\uparrow$ $Cu + 4HNO_3 \rightarrow Cu^{2+} + 2NO_3^- + 2H_2O + 2NO_2\uparrow$
e)	Carbon monoxide and carbon dioxide evolved (gas flammable with difficulty, calcium hydroxide solution turns milky)	$C_2O_4^{2-}$	$H_2SO_4 + C_2O_4^{2-} \rightarrow SO_4^{2-} + H_2O + CO\uparrow + CO_2\uparrow$

3.8 The Sodium Carbonate Extract Tests

To two spatula loads of the solid add five of anhydrous sodium carbonate (an excess) and $4 \, cm^3$ of water. Stir the mixture well and bring it to the boil. Centrifuge and divide the solution into six portions. For an explanation of any precipitates obtained refer to section 3.5.

Where an acid is added first in the tests below this must be done until the solution on stirring is acidic and the solution may then be warmed to facilitate the expulsion of carbon dioxide.

Test	Result	Inference
1 Add dilute nitric acid then silver nitrate solution.	White precipitate, soluble in ammonia solution.	Cl^-
	Cream precipitate, partially soluble in ammonia solution.	Br^-
	Pale yellow precipitate, insoluble in ammonia solution.	I^-
	A further test is to warm the solid with manganese(IV) oxide and concentrated sulphuric acid in a fume cupboard: if a halide is present the halogen is evolved; chlorine – yellow-green, bromine – brown, iodine – violet.	
2 Add dilute sulphuric acid then fresh iron(II) sulphate solution.	Brown ring.	NO_2^-
If no brown ring at the previous stage check that the solution is cool and then carefully add concentrated sulphuric acid so that it forms a lower layer.	Brown ring (may be a temporary one). (Bromides and iodides may give a similar result hence the necessity of keeping to the order suggested).	NO_3^-
3 Add dilute hydrochloric acid then barium chloride solution.	White precipitate	SO_4^{2-}

Metal ions other than sodium, potassium or ammonium may be referred to as 'heavy' metal ions. If the ion is Na^+, K^+ or NH_4^+ there will not be a precipitate. A heavy metal ion gives a precipitate which may be of the metal carbonate, hydroxide or oxide and thus a solution containing sodium ions, excess carbonate ions and the unknown anion remains:
$$M^{II}A + CO_3^{2-} \rightleftharpoons M^{II}CO_3\downarrow + A^{2-}$$
However, side effects (see section 3.5) may lead to incomplete elimination of, for example, tin.
$$2H^+ + CO_3^{2-} \rightarrow H_2O + CO_2\uparrow$$
The acidification may be done with dilute acetic (ethanoic) acid and effervescence from its reaction with excess sodium carbonate is a check upon the first stage.

1   $Ag^+ + X^- \rightarrow AgX\downarrow$
$AgX + 2NH_3 \rightleftharpoons [Ag(NH_3)_2]^+ + X^-$
The solubility products of silver chloride, bromide and iodide are 2×10^{-10}, 5×10^{-13} and $8 \times 10^{-17} \, mol^2 \, dm^{-6}$ respectively and only in the case of the bromide is a clear cut distinction between soluble or insoluble not obtained when ammonia is added.
$$2X^- + 4H_2SO_4 + MnO_2 \rightarrow Mn^{2+} + 4HSO_4^- + 2H_2O + X_2\uparrow$$

2   $2H^+ + 2NO_2^- \rightarrow H_2O + NO\uparrow + NO_2\uparrow$
$[Fe(H_2O)_6]^{2+} + NO \rightarrow [Fe(H_2O)_5NO]^{2+} \, (brown) + H_2O$
$NO_3^- + 3Fe^{2+} + 4H^+ \rightarrow 3Fe^{3+} + 2H_2O + NO\uparrow$
For bromides and iodides with concentrated sulphuric acid see section 3.7

3   $Ba^{2+} + SO_4^{2-} \rightarrow BaSO_4\downarrow$

Test	Result	Inference
4 Add dilute nitric acid, then 1 cm³ concentrated nitric acid and 3 cm³ ammonium molybdate [molybdate(VI)] solution.	Cold — blue coloration	Sn^{2+}
	Warm — yellow precipitate (if previously blue add more concentrated nitric acid)	PO_4^{3-}
	{Boil — yellow precipitate	AsO_4^{3-}}
5 Add dilute nitric acid then one drop of concentrated ammonia solution (the solution should now smell of ammonia), heat the solution and puff air through it until the smell has gone. This should yield a neutral solution and silver nitrate is now added.	White to pale yellow precipitate	Cl^-, Br^-, I^-, SO_3^{2-}, $S_2O_3^{2-}$, $C_2O_4^{2-}$, (SO_4^{2-}), (CH_3COO^-)
	Yellow precipitate	PO_4^{3-}, {AsO_3^{3-}}
	{Red precipitate	CrO_4^{2-}, AsO_4^{3-}}
	Black precipitate	S^{2-}, (Sn^{2+})
	N.B. Of these precipitates only the silver halide ones are insoluble if concentrated nitric acid is added.	
6 Add dilute sulphuric acid then potassium permanganate [manganate(VII)] solutions. If it is suspected that a metallic ion has decolorized the permanganate add some more permanganate.	Cold — goes colourless	NO_2^-, SO_3^{2-}, $S_2O_3^{2-}$, (Cl^-), Br^-, I^-, {AsO_3^{3-}}, Sn^{2+}, (Fe^{2+})
	Warm — goes colourless	$C_2O_4^{2-}$

4 Ammonium molybdate [molybdate(VI)] has the formula $(NH_4)_6Mo_7O_{24}\cdot4H_2O$ and the yellow precipitate given by a phosphate is $(NH_4)_2H(PMo_{12}O_{40})$. If the precipitate is washed with dilute ammonium nitrate solution it becomes the normal salt. The blue colloid is possibly Mo_2O_5. {An arsenate(V) yields a salt similar to that given by a phosphate but an arsenate(III) does not react.}

5   $Ag^+ + X^- \rightarrow AgX\downarrow$
  ($X^- = Cl^-$, Br^-, I^- or CH_3COO^-)

   $2Ag^+ + Y^{2-} \rightarrow Ag_2Y\downarrow$
  ($Y^{2-} = SO_3^{2-}$, $S_2O_3^{2-}$, $C_2O_4^{2-}$, SO_4^{2-} in concentrated solution, {CrO_4^{2-}})
   $3Ag^+ + PO_4^{3-} \rightarrow Ag_3PO_4\downarrow$
  {or AsO_3^{3-} or AsO_4^{3-}}
If an alkaline solution of the sodium carbonate extract is used by mistake then a pale yellow precipitate of silver carbonate will be obtained.
  $2Ag^+ + CO_3^{2-} \rightarrow Ag_2CO_3\downarrow$

If tin(II) ions are present then a redox reaction will occur yielding a black precipitate.
$2Ag^+ + Sn^{2+} \rightarrow 2Ag\downarrow + Sn^{4+}$
Silver salts of weak acids are soluble in nitric acid.
  $AgY \rightleftharpoons Ag^+ + Y^-$
$H^+ + Y^- \rightleftharpoons HY$

6   $2MnO_4^- + 5NO_2^- + 6H^+ \rightarrow 2Mn^{2+} + 3H_2O + 5NO_3^-$
    (or SO_3^{2-},   (or SO_4^{2-},
    or {AsO_3^{3-}})   or {AsO_4^{3-}})

$2MnO_4^- + 5S_2O_3^{2-} + 6H^+ \rightarrow 5S\downarrow + 2Mn^{2+} + 3H_2O + 5SO_4^{2-}$
$2MnO_4^- + 5Sn^{2+} + 16H^+ \rightarrow 2Mn^{2+} + 8H_2O + 5Sn^{4+}$
    (or {AsO_3^{3-}}   (or {AsO_4^{3-}}
    or Fe^{2+}, X^-)   or Fe^{3+}, X_2)
$2MnO_4^- + 5C_2O_4^{2-} + 16H^+ \rightarrow 2Mn^{2+} + 8H_2O + 10CO_2\uparrow$

3.9 The Cation Classification (Separation)

Obtain a solution of the substance (see section 3.2) and
test it as follows:

Test	Result	Inference
		The cation is in analytical group:
1 Add three drops of dilute hydro-chloric acid, then cool the solution.	A white precipitate may form: Hg_2Cl_2, AgCl or $PbCl_2$. If no precipitate keep the solution for (2).	I (proceed to section 3.11)
2 If the solution is of pH >4, reduce it to this by adding dilute hydrochloric acid. In a fume cupboard test a portion by adding two drops of hydrogen sulphide solution; if any coloured precipitate forms add hydrogen sulphide to all the solution.	A precipitate may form: black or brown HgS, PbS, Bi_2S_3, CuS, SnS orange Sb_4S_6 yellow SnS_2, {As_4S_6}, CdS white S (discard it). If no precipitate check test portion by diluting three times; if still none pour test portion away in fume cupboard. If no precipitate keep the main solution for (3).	II (proceed to section 3.12)

1   The solution must be cold because lead(II) chloride is soluble in hot water; dilute hydrochloric acid must be used because lead(II) chloride is soluble in the concentrated acid yielding tetrachloroplumbate(II) ions $[PbCl_4]^{2-}$

   The solubility products of these chlorides are Hg_2Cl_2 2×10^{-18} mol^3 dm^{-9}; AgCl 2×10^{-10} mol^2 dm^{-6}. $PbCl_2$ 2×10^{-5} mol^3 dm^{-9}.

2   A solution of hydrogen sulphide in acetone (propanone) containing 0.5% of water may be found to be more convenient than the gas obtained from a Kipp's apparatus. The control of acidity, which in turn controls the ionization of the hydrogen sulphide, is vital to achieve selective precipitation of sulphides.

   The solubility products of these sulphides are HgS 2×10^{-52} mol^2 dm^{-6}; PbS 1×10^{-28} mol^2 dm^{-6}; Bi_2S_3 2×10^{-72} mol^5 dm^{-15}; CuS 6×10^{-36} mol^2 dm^{-6}; SnS 1×10^{-26} mol^2 dm^{-6}; Sb_2S_3 1×10^{-93} mol^5 dm^{-15}; SnS_2 unknown; {As_4S_6 unknown}; CdS 8×10^{-27} mol^2 dm^{-6}

   Intermediate stages involving $PbS \cdot PbCl_2$ (red) or $HgCl_2 \cdot 2HgS$ (white) may occur in the precipitation.

   If oxidizing agents are present a white precipitate of sulphur or involving sulphate ions may form.
$$S^{2-} \rightarrow S\downarrow + 2e^-$$
$$S^{2-} + 4H_2O \rightarrow SO_4^{2-} + 8H^+ + 8e^-$$
The formation of a sulphate may lead to loss of Group V.
$$E^{2+} + SO_4^{2-} \rightarrow ESO_4\downarrow \text{ (white)}$$
Side effects include the reduction of iron(III) [and chromate] salts.
$$2Fe^{3+} + S^{2-} \rightarrow S\downarrow + 2Fe^{2+}$$
$$\{2CrO_4^{2-} + 3S^{2-} + 16H^+ \rightarrow 3S\downarrow + 2Cr^{3+} + 8H_2O\}$$

Test	Result	Inference
		The cation is in analytical group:
3 Take the remainder of the solution, add concentrated nitric acid and boil vigorously. Add two spatula loads of solid ammonium chloride and then concentrated ammonia solution dropwise until the solution is alkaline on stirring.	A precipitate may form: white $Al(OH)_3$ brown $Fe(OH)_3$ mauve or green $Cr(OH)_3$ off-white $Mn(OH)_2$. If no precipitate keep the solution for (4).	III (proceed to section 3.14)
4 Test a portion of the alkaline solution from (3) by adding hydrogen sulphide solution; if any precipitate forms add hydrogen sulphide to all the solution.	A precipitate may form: off-white ZnS buff MnS black NiS, CoS If no precipitate keep the main solution for (5).	IV (proceed to section 3.15)
5 To the solution from (4) add 0.5 cm^3 of concentrated ammonia solution and 1 cm^3 ammonium carbonate solution. Wait five minutes stirring occasionally.	A white precipitate may form: $CaCO_3$, $BaCO_3$, $\{SrCO_3\}$. If no precipitate keep the solution for (6).	V (proceed to section 3.16)
6 If the five tests described above have not yielded a precipitate then, unless the cation was ammonium (as found in section 3.5), the cation may be Mg^{2+}, K^+ or Na^+.		VI (proceed to section 3.17)

3 If concentrated nitric acid is added to a solution containing hydrogen sulphide then sulphur is precipitated.
$$8H^+ + 2NO_3^- + 3S^{2-} \rightarrow 3S\downarrow + 4H_2O + 2NO\uparrow$$
The vapours should be tested with lead acetate (ethanoate) solution on filter-paper.
$$Pb^{2+} + S^{2-} \rightarrow PbS\downarrow \text{ (black)}$$
The addition of concentrated nitric acid (a few drops) is vital if an iron salt is present to ensure that it is all in the iron(III) condition.
$$3Fe^{2+} + NO_3^- + 4H^+ \rightarrow 3Fe^{3+} + 2H_2O + NO\uparrow$$
If iron(II) ions are present then the precipitation may be incomplete.
$$Fe(OH)_2 + 2NH_4Cl \rightleftharpoons FeCl_2 + 2H_2O + 2NH_3$$
Another reason for adding concentrated nitric acid is that if an organic anion has been found then it must be destroyed by oxidation before the solution is made alkaline (0.5 cm^3 of concentrated nitric acid may be required). It is essential to add ammonium chloride as well as ammonia so that the concentration of hydroxide ions is kept low thus achieving a selective precipitation of hydroxides.

If the ammonia is contaminated with ammonium carbonate then there may be a loss of Group V cations.
$$E^{2+} + CO_3^{2-} \rightarrow ECO_3\downarrow \text{ (white)}$$
The solubility products of these hydroxides are $Al(OH)_3$ 1×10^{-32} mol^4 dm^{-12}; $Fe(OH)_3$ 8×10^{-40} mol^4 dm^{-12}; $Cr(OH)_3$ 1×10^{-30} mol^4 dm^{-12}; $Mn(OH)_2$ 4×10^{-14} mol^3 dm^{-9}.

The hydroxides are gelatinous precipitates and in analysis of mixtures Group V cations may be lost by adsorption. Manganese(II) hydroxide is white initially but darkens on oxidation by the air.
$$2Mn(OH)_2 + O_2 \rightarrow 2MnO_2 \cdot H_2O \text{ (brown)}$$
There are several cations that form complex ions under the circumstances and hence do not occur in this group, e.g. $[Zn(NH_3)_4]^{2+}$, $[MgCl_3]^-$, $[Mg(NH_3)_6]^{2+}$ (all colourless) and $[Ni(NH_3)_6]^{2+}$ (blue).

4 The addition of hydrogen sulphide to the alkaline solution precipitates the remaining insoluble sulphides. The solubility products (mol^2 dm^{-6}) of these sulphides are ZnS 2×10^{-24}; MnS 1×10^{-15}; NiS 2×10^{-26}; CoS 3×10^{-26}.

Nickel sulphide may not be precipitated if there is too much ammonium chloride present, when a brown solution may be seen, so some ammonia should then be boiled off. No coloured ions are left after this stage.

5 The solubility products (mol^2 dm^{-6}) of these carbonates are $CaCO_3$ 5×10^{-9}; $BaCO_3$ 6×10^{-10}; $\{SrCO_3$ $1 \times 10^{-10}\}$.

The precipitation may be slow. The solubility product of magnesium carbonate (1×10^{-5} mol^2 dm^{-6}) is not reached in this buffer solution.

{3.10 The Phosphate Separation

This is a necessary step if cations in Groups IV to VI allied with phosphate ions are to be found other than, if possible, by flame tests, and must be done when the nature of the solution is changed from acidic to alkaline in the Group III procedure. The presence of a phosphate will be known from section 3.8.

After boiling the solution with concentrated nitric acid add a few drops of zirconyl nitrate [zirconium(IV) nitrate oxide, $ZrO(NO_3)_2$] solution, stir and leave for three minutes before centrifuging. Continue by adding the zirconium salt solution dropwise until no more precipitation occurs.
$$ZrO^{2+} + PO_4^{3-} + H^+ \rightarrow ZrO(HPO_4)\downarrow \text{ (or } ZrO(H_2PO_4)_2\downarrow)$$
Discard the precipitates and proceed with Group III. The Group III precipitate contains (or is) zirconium(IV) hydroxide (white): this is not soluble in sodium hydroxide solution and does not interfere with the tests for the other members.}

3.11 The Group I Examination

1 Wash the precipitate twice with cold water. Add $4\,cm^3$ of water, stir and heat: lead(II) chloride is soluble in hot water. Centrifuge if a residue is obtained and proceed to (3).

---

2 Allow the solution to cool: fine white crystals forming indicate lead(II) chloride. Confirm the presence of lead by adding $1\,cm^3$ of potassium chromate [chromate(VI)] solution: a yellow precipitate indicates lead(II) chromate.
$$Pb^{2+} + CrO_4^{2-} \rightarrow PbCrO_4\downarrow$$

---

3 To the residue from (1) add concentrated ammonia solution dropwise, stir and warm. A black residue indicates the presence of mercury and mercury(II) aminochloride, silver chloride being soluble in ammonia solution.
$$AgCl + 2NH_3 \rightarrow [Ag(NH_3)_2]^+ + Cl^-$$
$$Hg_2Cl_2 + 2NH_3 \rightarrow Hg\downarrow + Hg(NH_2)Cl\downarrow + NH_4^+ + Cl^-$$

---

4 Acidify the solution by adding dilute nitric acid: the white precipitate which may form is of silver chloride; it may darken in light.
$$[Ag(NH_3)_2]^+ + Cl^- + 2H^+ \rightarrow AgCl\downarrow + 2NH_4^+$$
$$AgCl + h\nu \rightarrow \underset{\text{Black}}{Ag\downarrow} + \tfrac{1}{2}Cl_2$$

See section 3.18 for the Group I confirmatory tests.

3.12 Group II and the Examination of Group IIA

1 Wash the precipitate twice with hot water. Add $4\,cm^3$ of lithium hydroxide solution (containing potassium nitrate to prevent the formation of colloidal suspensions), stir and warm. Centrifuge if a residue forms and proceed to (2); if a solution is obtained the cation is considered under Group IIB—proceed to section 3.13.

Group IIB sulphides dissolve in the alkaline solution.
$$2Sb_2S_3 + 4OH^- \rightarrow 3SbS_2^- + SbO_2^- + 2H_2O$$
Thioantimo-
nate(III)

$\{As_2S_3$ reacts likewise; $As_2S_5 \rightarrow$ thioarsenates(V)$\}$
$$SnS + 4OH^- \rightarrow SnO_2^{2-} + S^{2-} + 2H_2O$$

Tin(II) sulphide dissolves slowly at the boiling-point of the solution to give a stannate(II) which is, however, readily oxidized to a stannate(IV): the dissolution may be facilitated by the addition of yellow ammonium sulphide solution.
$$SnS + S_2^{2-} \rightarrow SnS_3^{2-}$$
Thiostannate(IV)

Thus the initial presence of a tin(II) salt may lead to the eventual discovery of a tin(IV) salt in Group IIB.
$$SnS_2 + 6OH^- \rightarrow SnO_3^{2-} + 2S^{2-} + 3H_2O$$

---

2 Warm the residue with $2\,cm^3$ of water and $2\,cm^3$ of concentrated nitric acid. A black residue indicates the presence of mercury(II) sulphide. The other sulphides dissolve.
$$3PbS + 8H^+ + 2NO_3^- \rightarrow 3S\downarrow + 3Pb^{2+} + 4H_2O + 2NO\uparrow$$
CuS and CdS likewise
$$Bi_2S_3 + 6H^+ \rightarrow 2Bi^{3+} + 3H_2S\uparrow$$
There may be some sulphate formation.
$$3CuS + 8H^+ + 8NO_3^- \rightarrow 3Cu^{2+} + 3SO_4^{2-} + 4H_2O + 8NO\uparrow$$
This may lead to early precipitation of lead ions.
$$Pb^{2+} + SO_4^{2-} \rightarrow PbSO_4\downarrow \text{ (white)}$$
If a white suspension forms and it is suspected to be sulphur it may be helpful to add two spatula loads of ammonium chloride and boil the solution to cause coagulation.

---

3 To the solution from (2) add $2\,cm^3$ of dilute sulphuric acid. A white precipitate indicates the presence of lead(II) sulphate (equation above).

---

4 To the solution from (3) add a concentrated solution of ammonia until the solution is alkaline on stirring. A white precipitate indicates the presence of bismuth(III) hydroxide.
$$Bi^{3+} + 3OH^- \rightarrow Bi(OH)_3\downarrow$$

---

5 If the solution from (4) is blue it indicates the presence of tetraamminecopper(II) ions—look vertically down the test-tube. $\{$If the solution is colourless add hydrogen sulphide solution: a yellow precipitate indicates the presence of cadmium(II) sulphide$\}$.

$$Cu^{2+} + 2OH^- \rightarrow Cu(OH)_2\downarrow \text{ (pale blue)}$$
$$Cu^{2+} + 4NH_3 \rightarrow [Cu(NH_3)_4(H_2O)_2]^{2+} \text{ (deep blue)}$$
$$Cd^{2+} + 2OH^- \rightarrow Cd(OH)_2\downarrow \text{ (white)}$$
$$Cd^{2+} + 4NH_3 \rightarrow [Cd(NH_3)_4]^{2+} \text{ (colourless)}$$
$$Cd^{2+} + S^{2-} \rightarrow CdS\downarrow$$
See section 3.19 for the Group IIA confirmatory tests.

3.13 The Group IIB Examination

1 Acidify the solution obtained from section 3.12(1) with dilute hydrochloric acid to regain the precipitate which must be of a Group IIB element. If a white precipitate appears it is probably of sulphur and should be discarded.
$$3SbS_2^- + SbO_2^- + 4H^+ \rightarrow 2Sb_2S_3\downarrow + 2H_2O$$
Orange

$\{AsS_2^-$ reacts likewise$\}$
$$SnS_3^{2-} + 2H^+ \rightarrow SnS_2\downarrow + H_2S$$
$$SnO_3^{2-} + 2S^{2-} + 6H^+ \rightarrow SnS_2\downarrow + 3H_2O$$
Yellow

$\{$If arsenic may be present the precipitate should be boiled with $4\,cm^3$ of ammonium carbonate solution: arsenic(III) sulphide would dissolve and its return after acidification with dilute hydrochloric acid and the addition of hydrogen sulphide solution would be a strong indication of the presence of arsenic.
$$2As_2S_3 + 2CO_3^{2-} \rightarrow 3AsS_2^- + AsO_2^- + 2CO_2\uparrow$$
$$AsS_2^- + 4H^+ \rightarrow As^{3+} + 2H_2S\uparrow$$
AsO_2^- reacts likewise
$$2As^{3+} + 3S^{2-} \rightarrow As_2S_3\downarrow \}$$
Yellow

---

2 Redissolve the precipitate from (1) in hot concentrated hydrochloric acid and divide it into two portions.
$$SnS_2 + 4H^+ \rightarrow Sn^{4+} + 2H_2S\uparrow$$
$$Sb_2S_3 + 6H^+ \rightarrow 2Sb^{3+} + 3H_2S\uparrow$$
a) Dilute the solution with an equal volume of water and then add hydrogen sulphide solution: the reappearance of an orange precipitate indicates antimony(III) sulphide.
$$2Sb^{3+} + 3S^{2-} \rightarrow Sb_2S_3\downarrow$$

b) Add a piece of aluminium foil or clean iron wire, boil the mixture and then, after allowing it to cool, add mercury(II) chloride solution: a white precipitate of mercury(I) chloride turning to a grey precipitate of mercury indicates the presence of tin(II) ions. The original substance must be studied to establish whether it is a tin(II) or a tin(IV) salt, see section 3.20.
$$Fe + Sn^{4+} \rightarrow Sn^{2+} + Fe^{2+}$$
$$Sn^{2+} + 2Hg^{2+} + 2Cl^- \rightarrow Hg_2Cl_2\downarrow + Sn^{4+}$$
$$Hg_2Cl_2 + Sn^{2+} \rightarrow 2Hg\downarrow + Sn^{4+} + 2Cl^-$$
See section 3.20 for the Group IIB confirmatory tests.

3.14 The Group III Examination

1 Wash the precipitate twice with hot water. Add $2\,cm^3$ dilute sodium hydroxide solution and $2\,cm^3$ hydrogen peroxide solution and boil the mixture. A solution is obtained from chromium and aluminium compounds.

$$2Cr(OH)_3 + 3H_2O_2 + 4OH^- \rightarrow 2CrO_4^{2-} + 8H_2O$$
$$Al(OH)_3 + 3OH^- \rightarrow [Al(OH)_6]^{3-}$$
$$\text{Some } 2H_2O_2 \rightarrow 2H_2O + O_2\uparrow$$

A brown residue indicates the presence of iron(III) hydroxide: dissolve a portion of it in dilute hydrochloric acid and add potassium hexacyanoferrate(II); a precipitate of Prussian blue confirms the presence of iron. The original substance must be studied to establish whether it is an iron(II) or an iron(III) salt, (see section 3.21).

$$Fe(OH)_3 + 3H^+ \rightarrow Fe^{3+} + 3H_2O$$
$$4Fe^{3+} + 3[Fe(CN)_6]^{4-} + 14H_2O \rightarrow Fe_4[Fe(CN)_6]_3 \cdot 14H_2O$$

If the Prussian blue precipitate does not form it is possible that manganese(II) hydroxide has appeared in Group III. Dissolve the remainder of the precipitate in $2\,cm^3$ of dilute nitric acid, add sodium bismuthate(V), stir and warm before centrifuging: a pink solution containing permanganate [manganate(VII)] ions indicates the original presence of manganese ions.

$$Mn(OH)_2 + 2H^+ \rightarrow Mn^{2+} + 2H_2O$$
$$2Mn^{2+} + 5BiO_3^- + 14H^+ \rightarrow 2MnO_4^- + 5Bi^{3+} + 7H_2O$$

2 Divide the solution from (1) into two portions.
a) Acidify the solution with dilute acetic (ethanoic) acid and then add two drops of lead(II) acetate (ethanoate) solution: a yellow precipitate of lead(II) chromate [chromate(VI)] indicates the presence of chromium in the original salt.

$$Pb^{2+} + CrO_4^{2-} \rightarrow PbCrO_4\downarrow$$

b) Acidify the solution with dilute hydrochloric acid, then add just one drop of litmus solution and make the solution just alkaline with dilute ammonia solution. Centrifuge: a white precipitate which adsorbs the dye forming a 'blue lake' indicates the presence of aluminium ions.

$$[Al(OH)_6]^{3-} + 6H^+ \rightarrow Al^{3+} + 6H_2O$$
$$Al^{3+} + 3OH^- \rightarrow Al(OH)_3\downarrow$$

See section 3.21 for the Group III confirmatory tests.

3.15 The Group IV Examination

1 Wash the precipitate twice with hot water. Add $1\,cm^3$ of water and $2\,cm^3$ of dilute hydrochloric acid. A solution is obtained from zinc and manganese compounds.

$$ZnS + 2H^+ \rightarrow Zn^{2+} + H_2S\uparrow$$

MnS reacts likewise
A black residue indicates the presence of nickel or cobalt sulphide.

2 To the solution from (1) add excess sodium hydroxide solution. An off-white residue indicates the presence of manganese(II) hydroxide oxidizing in the air.

$$Zn^{2+} + 2OH^- \rightarrow Zn(OH)_2\downarrow$$

Mn^{2+} reacts likewise   White

$$Zn(OH)_2 + 2OH^- \rightarrow [Zn(OH)_4]^{2-}$$
$$2Mn(OH)_2 + O_2 \rightarrow 2MnO_2 \cdot H_2O$$
$$\text{Brown}$$

3 To the solution from (2) add hydrogen sulphide solution. An off-white residue indicates the presence of zinc sulphide.

$$[Zn(OH)_4]^{2-} + 4H^+ + S^{2-} \rightarrow ZnS\downarrow + 4H_2O$$

See section 3.22 for the Group IV confirmatory tests.

3.16 The Group V Examination

1 Wash the precipitate twice with hot water. Dissolve it by warming with the minimum quantity of dilute acetic (ethanoic) acid. To a portion of the solution add two drops of potassium chromate [chromate(VI)] solution and allow it to stand: if there is a yellow precipitate barium ions are present.

$2H^+ + ECO_3 \rightarrow E^{2+} + H_2O + CO_2 \uparrow$
$Ba^{2+} + CrO_4^{2-} \rightarrow BaCrO_4 \downarrow$ (yellow)

$\{$The solution is acidic and the solubility product of strontium chromate is not exceeded.

$2CrO_4^{2-} + 2H^+ \rightleftharpoons Cr_2O_7^{2-} + H_2O \}$

If there is no precipitate proceed to (3) if strontium is not in the syllabus or to (2) if strontium is in the syllabus. If a mixture is being analyzed all the barium ions must be precipitated as barium chromate at this stage.

2 $\{$Divide the solution from (1) into two portions.
a) Add saturated calcium sulphate solution: a white precipitate, which may be slight and slow in forming, indicates the presence of strontium ions.
$Sr^{2+} + SO_4^{2-} \rightarrow SrSO_4 \downarrow$
b) If a mixture is being analyzed and strontium has been found to be present add dilute sulphuric acid, discard the precipitate of strontium sulphate which forms together with a little calcium sulphate and then test for calcium ions as in (3).

The solubility product of calcium sulphate is $2 \times 10^{-4} \, mol^2 \, dm^{-6}$ whereas that of strontium sulphate is $4 \times 10^{-7} \, mol^2 \, dm^{-6}$. The dilute sulphuric acid removes most of the calcium ions as well as most of the strontium ions.$\}$

3 To the solution from (1) add dilute ammonia solution and ammonium oxalate (ethanedioate) solution: a white precipitate indicates the presence of calcium ions.
$Ca^{2+} + C_2O_4^{2-} + H_2O \rightarrow CaC_2O_4 \cdot H_2O \downarrow$
See section 3.23 for the Group V confirmatory tests.

3.17 The Group VI Examination

1 $\{$If a mixture is being analyzed and calcium was found in Group V add ammonium oxalate (ethanedioate) and discard the precipitate of calcium oxalate-1-water.$\}$

2 Divide the solution into two portions.
a) Add dilute ammonia solution and disodium hydrogenphosphate solution and stand: a white precipitate indicates the presence of magnesium ions.
$Mg^{2+} + NH_4^+ + PO_4^{3-} + 6H_2O \rightarrow MgNH_4PO_4 \cdot 6H_2O \downarrow$
b) Put the solution in a crucible over a Bunsen burner in a fume cupboard and evaporate the solution to dryness, heating strongly until no more fumes are evolved. Do a flame test on the residue: this should confirm the result of the test done in section 3.4.
See section 3.24 for the Group VI confirmatory tests.

Confirmatory Tests

3.18 Group I

Lead(II) ions
With potassium iodide solution, a hot solution of lead chloride yields a yellow precipitate.
$$Pb^{2+} + 2I^- \rightarrow PbI_2\downarrow$$
If a small quantity of the precipitate is warmed with water a colourless solution is obtained from which 'golden spangles' form on cooling.

Silver(I) ions
With potassium iodide the white precipitate of silver chloride is transformed into a pale yellow precipitate of silver iodide.
$$AgCl + I^- \rightarrow AgI\downarrow + Cl^-$$

Mercury(I) ions
When the black precipitate of mercury and mercury aminochloride is warmed with a little concentrated nitric acid to make a solution (boil to destroy excess acid) and then potassium iodide solution added, a yellow precipitate which rapidly turns red is formed. Mercury(II) iodide has two allotropes with a transition temperature of 400 K (127 °C).
$$3Hg + 8H^+ + 2NO_3^- \rightarrow 3Hg^{2+} + 4H_2O + 2NO\uparrow$$
$$HgNH_2Cl + 2H^+ \rightarrow Hg^{2+} + NH_4^+ + Cl^-$$
$$Hg^{2+} + 2I^- \rightarrow HgI_2\downarrow$$
Check whether the original substance is a mercury(I) or (II) salt by adding to its solution potassium iodide solution: in the former case a yellow precipitate turning green is formed.
$$Hg_2^{2+} + 2I^- \rightarrow Hg_2I_2\downarrow$$

3.19 Group IIA

Mercury(II) ions
Treat the black precipitate of mercury(II) sulphide as in the previous section.

Check whether the original substance is a mercury(I) or (II) salt as in the previous section.

Lead(II) ions
Lead sulphate is soluble in concentrated hydrochloric acid and on adding potassium iodide solution a yellow precipitate forms.
$$PbSO_4 + 4Cl^- \rightarrow PbCl_4^{2-} + SO_4^{2-}$$
$$PbCl_4^{2-} + 2I^- \rightarrow PbI_2\downarrow + 4Cl^-$$

Bismuth(III) ions
Dissolve the bismuth hydroxide in the minimum quantity of concentrated hydrochloric acid and then pour the solution into cold water: a white precipitate reappears.
$$Bi(OH)_3 + 3H^+ \rightarrow Bi^{3+} + 3H_2O$$
$$Bi^{3+} + Cl^- + H_2O \rightleftharpoons BiOCl\downarrow + 2H^+$$
A second test for bismuth (Kobell 1871) is to heat gently the bismuth compound with a little charcoal, sulphur and potassium iodide near the end of a carbon block by means of a Bunsen burner and a blow-pipe. The red compound bismuth(III) iodide volatilizes and condenses nearby.
$$2Bi_2O_3 + 3C + 6S + 12KI \rightarrow 4BiI_3 + 6K_2S + 3CO_2\uparrow$$

Copper(II) ions
To the blue solution add dilute acetic (ethanoic) acid and then potassium hexacyanoferrate(II) solution: a brown precipitate forms.
$$[Cu(NH_3)_4]^{2+} + 4H^+ \rightarrow Cu^{2+} + 4NH_4^+$$
$$2Cu^{2+} + [Fe(CN)_6]^{4-} \rightarrow Cu_2Fe(CN)_6\downarrow$$

Cadmium(II) ions
Dissolve the cadmium sulphide in concentrated hydrochloric acid; divide the solution into two portions.
a) Add sodium hydroxide solution slowly until it is present in excess: a white precipitate forms.
$$Cd^{2+} + 2OH^- \rightarrow Cd(OH)_2\downarrow$$
b) Add ammonia solution slowly until it is present in excess: a white precipitate forms (equation above) and then dissolves.
$$Cd(OH)_2 + 4NH_3 \rightarrow [Cd(NH_3)_4]^{2+} + 2OH^-]$$

3.20 Group IIB

Antimony(III) compounds
Perform the Marsh test in a fume cupboard. Put the compound, granulated zinc and concentrated hydrochloric acid in a test-tube fitted with a jet through the stopper. Collect the hydrogen evolved in a ignition tube and when it burns quietly ignite it at the jet; allow the flame to impinge on the cold exterior surface of a crucible: a black stain is produced which is insoluble in sodium hypochlorite [chlorate(I)] solution.
$$Zn + 2H^+ \rightarrow Zn^{2+} + H_2\uparrow$$
$$3Zn + Sb^{3+} + 3H^+ \rightarrow 3Zn^{2+} + SbH_3\uparrow$$
$$2SbH_3 \rightarrow 2Sb\downarrow + 3H_2$$
$$2H_2 + O_2 \rightarrow 2H_2O$$

Arsenic(III) compounds
In the Marsh test (described above) a black stain is produced which is soluble in the hypochlorite solution.
$$2As + 5OCl^- + 3H_2O \rightarrow 2AsO_4^{3-} + 5Cl^- + 6H^+$$

Tin(II) ions
The original precipitate in Group II should have been brown and the original substance in solution should on treatment with mercury(II) chloride solution give a white precipitate which slowly turns grey.
$$2Hg^{2+} + Sn^{2+} + 2Cl^- \rightarrow Hg_2Cl_2\downarrow + Sn^{4+}$$
$$Hg_2Cl_2 + Sn^{2+} \rightarrow 2Hg\downarrow + Sn^{4+} + 2Cl^-$$

Tin(IV) ions
The original precipitate in Group II should have been yellow and the original substance in solution does not react with mercury(II) chloride solution.

Iron compounds

The original substance must be studied in the form supplied and in solution to ascertain whether it is an iron(II) or (III) compound.

Test	Result if Iron(II)	Result if Iron(III)
1 Colour	Green	Yellow-brown
2 Addition of sodium hydroxide or ammonia solution	White, rapidly turning green, gelatinous precipitate	Red-brown gelatinous precipitate
3 Addition of potassium thio-cyanate solution	Colourless (but may be slightly red)	Red coloration
4 Addition of potassium hexacyanoferrate(II) solution	White precipitate which rapidly turns blue	Prussian blue precipitate and coloration
5 Addition of potassium hexacyanoferrate(III) solution	Prussian blue precipitate and coloration	Brown or Berlin green coloration

2
$$Fe^{2+} + 2OH^- \rightarrow Fe(OH)_2 \downarrow \text{ (white)}$$
$$6Fe(OH)_2 + 2H_2O + O_2 \rightarrow 2Fe_3(OH)_8 \downarrow \text{ (green)}$$
$$Fe^{3+} + 3OH^- \rightarrow Fe(OH)_3 \downarrow \text{ (red-brown)}$$

3 $Fe^{3+} + SCN^- \rightarrow FeSCN^{2+}$ (red)

4 $2K^+ + Fe^{2+} + [Fe^{II}(CN)_6]^{4-} \rightarrow (K^+)_2 [Fe_2^{II}(CN)_6]^{2-} \downarrow$
(white)

5 either $4Fe^{3+} + 3[Fe(CN)_6]^{4-} + 14H_2O$
or $3Fe^{2+} + Fe^{3+} + 3[Fe(CN)_6]^{3-} + 14H_2O$
gives $Fe_4[Fe(CN)_6]_3 \cdot 14H_2O$ (Prussian blue)

Chromium(III) ions

Obtain a solution of a chromate [chromate(VI)] as in section 3.14(1). To the solution of the chromate add dilute sulphuric acid and then excess hydrogen peroxide: the yellow solution becomes orange on acidification then turns blue and finally green. The blue coloured compound is soluble in, and stabilized by, diethyl ether (ethoxyethane). **Care** – flammable.
$$2CrO_4^{2-} + 2H^+ \rightarrow Cr_2O_7^{2-} + H_2O$$
$$Cr_2O_7^{2-} + 4H_2O_2 + 2H^+ \rightarrow 2CrO_5 + 5H_2O$$
$$2CrO_5 + 6H^+ \rightarrow 2Cr^{3+} + 2H_2O + H_2O_2 + 3O_2 \uparrow$$
CrO_5 is chromium oxide diperoxide and the structure could be written $Cr^{VI}O_5$ or $CrO(O_2)_2$.

Aluminium(III) ions

Acidify the solution containing aluminium ions with dilute acetic (ethanoic) acid, add alizarin

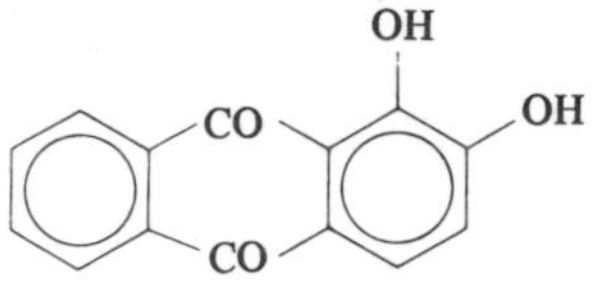

and then ammonia solution until the solution is alkaline: a red 'lake' forms, i.e. the dye is adsorbed on the precipitate.
$$Al^{3+} + 3OH^- \rightarrow Al(OH)_3 \downarrow$$

An alternative test is to heat the aluminium compound on a carbon block; a white residue is obtained which if moistened with one drop of cobalt(II) nitrate solution and then reheated gives a blue residue (Thenard's blue, $CoO \cdot Al_2O_3$).

3.22 Group IV

Cobalt(II) ions

Cobalt compounds (which are usually pink) give a blue borax bead: borax [disodium tetraborate(III)-10-water] is heated on a platinum wire until it has given a colourless bead, then one small crystal of the substance is reheated with the bead.

Alternatively dissolve the cobalt sulphide precipitate in concentrated nitric acid and boil the solution until it is nearly dry. Then divide into two portions.

$$CoS + 9HNO_3 + H^+ \rightarrow Co^{3+} + SO_4^{2-} + 5H_2O + 9NO_2\uparrow$$

a) Add two drops of dilute hydrochloric acid, $2\,cm^3$ potassium nitrite [nitrate(III)] solution and finally concentrated potassium chloride solution: a yellow precipitate forms.

$$3K^+ + Co^{3+} + 6NO_2^- + 3H_2O \rightarrow (K^+)_3[Co(NO_2)_6]^{3-}\cdot3H_2O\downarrow$$

b) Add 'nitroso-R-salt', i.e. disodium 1-nitroso-2-hydroxynaphthalene-3, 6-disulphonate: a red precipitate forms.

Nickel(II) ions

Nickel compounds (which are usually green) give a brown borax bead—see above.

Dissolve the precipitate of nickel sulphide (as above) and then add two drops of dilute sulphuric acid, four drops of dimethylglyoxime (butanedione dioxime) and six drops of dilute ammonia solution: a red precipitate forms, warming perhaps being necessary.

Manganese(II) ions

The fact that manganese(II) hydroxide darkens by atmospheric oxidation was mentioned in section 3.15. The test with sodium bismuthate [bismuthate(V)] was described in section 3.14.

$$2Mn^{2+} + 5BiO_3^- + 14H^+ \rightarrow \underset{\text{Purple}}{2MnO_4^-} + 5Bi^{3+} + 7H_2O$$

Zinc(II) ions

Dissolve the zinc sulphide precipitate in dilute hydrochloric acid and then add potassium hexacyanoferrate(II) slowly until it is present in excess: a white precipitate forms and then dissolves.

$$ZnS + 2H^+ \rightarrow Zn^{2+} + H_2S\uparrow$$
$$3Zn^{2+} + 2K^+ + 2Fe(CN)_6^{4-} \rightarrow (K^+)_2(Zn^{2+})_3[Fe^{II}(CN)_6^{4-}]_2$$

An alternative test is to heat the zinc compound on a carbon block: a white residue is obtained which if moistened with one drop of cobalt(II) nitrate solution and then reheated gives a green residue (Rinmann's green, $CoO\cdot ZnO$).

Calcium(II) ions

The calcium sulphate precipitated is soluble in a concentrated solution of ammonium sulphate with the formation of $[Ca(SO_4)_2]^{2-}$ ions.

An alternative test which can be applied to a solution containing Ca^{2+} ions is to add potassium hexacyanoferrate(II) solution in excess and then to saturate the solution with ammonium chloride: a white precipitate forms which has the approximate composition $NH_4^+ K^+ Ca^{2+} [Fe(CN)_6]^{4-}$.

{Strontium(II) ions

A precipitate of strontium sulphate is insoluble in a concentrated solution of ammonium sulphate, thus serving to check the distinction from calcium.}

Barium(II) ions

Dissolve the yellow precipitate of barium chromate [chromate(VI)] in dilute hydrochloric acid and then add dilute sulphuric acid: a white precipitate appears.

$$BaCrO_4 + 2H^+ \rightarrow Ba^{2+} + H_2CrO_4$$
$$Ba^{2+} + SO_4^{2-} \rightarrow BaSO_4\downarrow$$

Magnesium(II) ions

Ammonium ions must be eliminated and then to the solution of the substance are added two drops of dilute hydrochloric acid, two drops of Magneson I or II and finally sodium hydroxide solution dropwise until the mixture is alkaline: a blue precipitate confirms the presence of magnesium ions. Magneson I is 4-nitrobenzeneazoresorcinol [4-nitrobenzeneazo — (2,4-dihydroxy)-benzene] and Magneson II is 4-nitrobenzeneazo-1-naphthol [4-nitrobenzeneazonaphth-1-ol]. See section 2.4(b).

An alternative test is to heat the magnesium compound on a carbon block: a white residue is obtained which if moistened with one drop of cobalt(II) nitrate solution and then reheated gives a pink residue (CoO·MgO).

Sodium(I) ions

There are no fully reliable tests, but two that may be tried using a concentrated solution of the substance are the following.

a) Add potassium hydroxide and a hot concentrated solution of potassium antimonate(V): a white precipitate appears having the formula $NaSb(OH)_6 \cdot H_2O$. It is sparingly soluble in boiling water.

b) Add zinc(II) uranyl(VI) acetate (or the magnesium or nickel salt) and dilute acetic (ethanoic) acid: a yellow precipitate appears having the formula $NaZn(UO_2)_3 \cdot 9CH_3COOH \cdot 9H_2O$.

Potassium(I) ions

There are no fully reliable tests, but one that may be tried using a concentrated solution of the substance from which ammonium ions must be absent is as follows.

Add sodium hexanitrocobaltate(III) solution and dilute acetic (ethanoic) acid: a yellow precipitate appears which has the formula $K_3Co(NO_2)_6 \cdot 3H_2O$. It may be slow in forming.

IV Organic Chemistry by Homologous Series

Introduction

Organic chemicals tend, in general, to be more dangerous than many of the inorganic chemicals used in schools. They should not be handled and they should not be smelled more than is necessary. A fume cupboard is frequently advised and safety-glasses should be worn in addition to laboratory coats.

The preparations are carried out on a small scale because they are only designed to illustrate processes. The tests likewise are to illustrate the reactions of members of a homologous series: again, from this aspect large quantities of material should not be employed. The fire hazard is greater than in inorganic chemistry. A typical basic set of apparatus for these practicals is that made by Quickfit (29 BU) and for the reactions ignition tubes (50 x 10 mm) and test-tubes (100 x 16 mm) may be used.

When you are required to set up the apparatus with a *reflux* condenser it should be clamped over a Bunsen burner, tripod and gauze as shown in figure 4.A. A few anti-bumping granules (sintered aluminium oxide or carborundum chips) put in the bottom of the 50 cm³ flask ensure that the formation of bubbles of vapour is reasonably smooth. The water in the condenser should flow upwards so that the condenser is kept full. The flask should never be more than half full. If a reagent

has to be added in portions this should be done by dropping it centrally down the condenser (see figure 4.B). Sometimes a beaker of water may be used as a water bath between the burner and the flask.

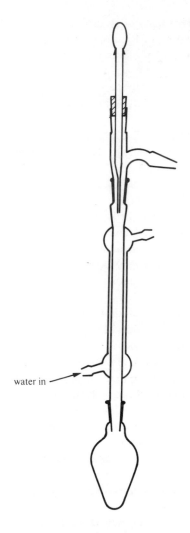

water in

Figure 4.B

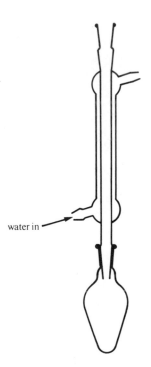

water in

Figure 4.A

To *distil* a substance the apparatus should be set up a as in figure 4.C. If a liquid with a boiling-point above 425 K (152 °C) is being distilled water should not be run through the condenser, which will then act as an air condenser. If a flammable solvent is being distilled off a water bath may be used and/or the vapours led into a conical flask. To the side-arm of the flask attach a piece of tubing leading vapours away from the flames. It may be useful to cool the conical flask as well; set up the apparatus as in figure 4.D.

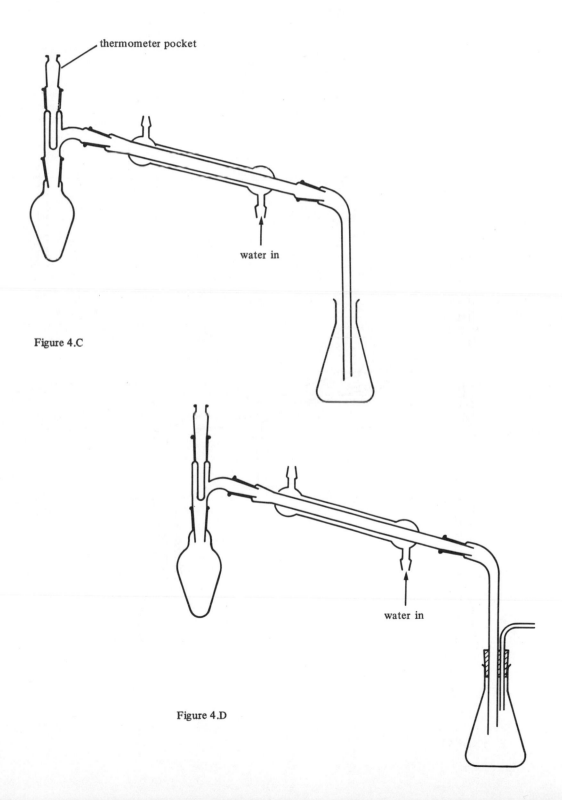

thermometer pocket

water in

Figure 4.C

water in

Figure 4.D

To *steam distil* a substance the apparatus should be set up as in figure 4.E. The steam is generated in a conical flask fitted with a safety tube. The steam is led into the pear-shaped flask by a glass tube through a cork: the tube should have a jet at the lower end.

For *fractional distillation* of a mixture the apparatus should be set up as in figure 4.F: water is not run through the vertical condenser.

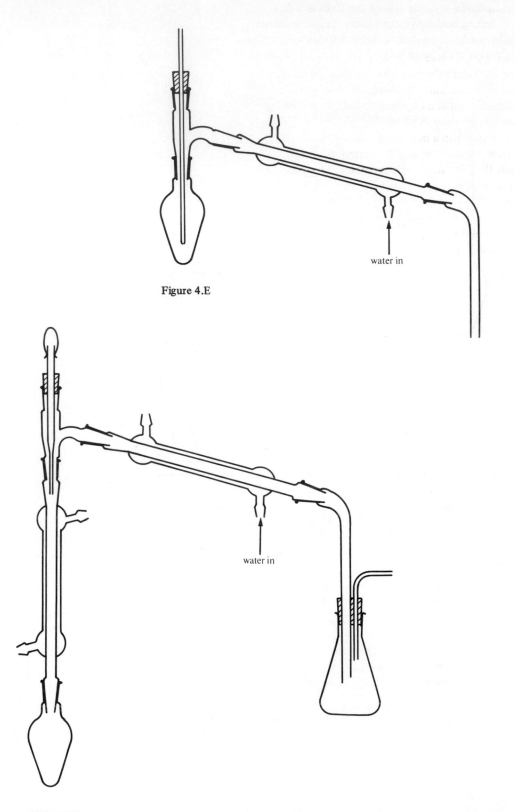

Figure 4.E

water in

Figure 4.F

water in

4.1 Melting- and Boiling-Points

Solids

The melting-point of a solid is a good guide to its purity and may be useful for identifying the solid. Usually a pure substance will melt over a range of 1 K (°C) or less.

1 Simple melting-points

The substance is crushed with the sealed end of a melting-point tube on a filter-paper or with a pestle in a clean mortar and the powdered solid placed to a depth of 1 cm in a capillary (melting-point) tube closed at one end. The tube is then placed alongside a thermometer in a beaker of silicone oil (e.g. MS 550) and the beaker then slowly heated. Surface tension holds the tube to the thermometer. The rate of heating is decreased as the expected melting-point is approached: the crystals then collapse to a clear liquid.

2 Mixed melting-points

This method has to be adopted in difficult cases of identification. The substance X is found to have a melting-point very close to those recorded in the reference books for A and B. Samples of A and B are studied by melting-point determination and then a little X added to each for two further determinations: if X is A then the melting-point of the mixture is unchanged whilst that of the mixture of X and B is lower than that of B alone.

Liquids

The boiling-point of a liquid is a good guide to its purity and may be useful for identifying the liquid. The determination must be done at a known pressure because boiling-point, unlike melting-point, is affected considerably by changes in pressure. A pure substance will boil over a range of 1 K (°C) or less.

The liquid should be heated in a flask and a thermometer held in the vapour when it boils. Then the thermometer is lowered into the liquid: the same value is recorded if the liquid is pure.

If the liquid is very flammable precautions must be taken to keep the vapours away from the Bunsen flame. It may be advisable to use a water bath and to set up the apparatus as for distillation (figure 4.C).

In difficult cases, e.g. decomposition occurring on boiling, the liquid must be converted into a solid derivative and the melting-point of that determined.

e.g. 4-methylbenzophenone
(4-methyldiphenylmethanone)
m.p. 332–3 K (59–60°C)
b.p. 599 K (326°C)

4.2 The Preparation and Properties of Methane

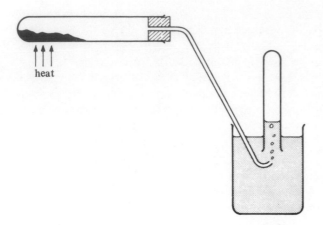

heat

1 Set up the apparatus as in the figure. Into the boiling tube put a well-ground mixture of roughly equal masses of anhydrous sodium acetate (ethanoate) and soda-lime (sodium hydroxide and calcium hydroxide). Heat the mixture and collect test-tubes of the gas evolved: cork the tubes.
Care: beware of sucking-back.

2 Test the methane as follows:
a) Add $2\,cm^3$ of bromine water, recork the test-tube and shake it.
b) Mix $1\,cm^3$ of potassium permanganate [manganate(VII)] solution and $1\,cm^3$ of sodium carbonate solution (Baeyer's reagent): add the mixture to the methane; recork the test-tube and shake it.
c) Apply a lighted splint to an open test-tube containing the gas. This should not be the first or second tube collected because they will also contain air.

Write equations for the reactions that you observe.

4.3 The Preparation of Hexane

This is done by Wurtz's method.

1 Set up the apparatus for refluxing (figure 4.A). Put $10\,cm^3$ of 1-bromopropane (13.5 g) in the flask and add 3 g of small pieces of sodium. Heat the flask gently for one hour: once the reaction has started it may not be necessary to continue heating. These quantities represent about 0.1 mole of each reagent.

2 Cool the flask in a beaker of water. Add $10\,cm^3$ of ethanol slowly down the condenser with constant shaking of the flask and then add $15\,cm^3$ of water likewise. These operations convert surplus sodium to sodium ethoxide and then to sodium hydroxide. Finally reflux the mixture for a few minutes to hydrolyze any bromopropane remaining.

3 Again cool the flask in a beaker of water. Transfer the contents of the flask to a separating funnel (about $100\,cm^3$ capacity). Run off the lower aqueous layer. Wash the upper layer three times with $20\,cm^3$ portions of water and then put the wet product in a 100×16 mm test-tube. Add two spatula loads of anhydrous magnesium sulphate and shake the tube for a few minutes. Separate the crystals by centrifuging the tube and put the clear product in a small flask set up for distillation (figure 4.C). Add an anti-bumping granule and carry out the distillation: collect the portion distilling at $342 \pm 2\,K$ ($69 \pm 2\,°C$). Hexane has a density of $0.66\,g/cm^3$. Write the equations for the reactions concerned. Calculate the percentage of the theoretical yield obtained.

4.4 The Properties of Alkanes

Pentane, hexane, heptane and cyclohexane are studied.

1 Shake 2 cm^3 of water with a few drops of the alkane. Test the water (a solution?) with pH paper.

2 Put a few drops of the alkane on a crucible lid and ignite it with a lighted splint.

3 Put 1 cm^3 of potassium permanganate [manganate(VII)] solution and 1 cm^3 of dilute sulphuric acid in a test-tube; add a few drops of the alkane and shake well.

4 Put 1 cm^3 of potassium permanganate solution and 1 cm^3 of sodium carbonate solution (Baeyer's reagent) in a test-tube, add a few drops of the alkane and shake well.

5 Put 1 cm^3 of bromine water in a test-tube; add a few drops of the alkane and shake well.

6 Put 0.5 cm^3 bromine dissolved in carbon tetrachloride (tetrachloromethane) in a test-tube, add a few drops of the alkane, observe, then wash away down the fume cupboard drain.

7 To 1 cm^3 concentrated nitric acid add 1 cm^3 concentrated (or fuming) sulphuric acid and then carefully add 0.5 cm^3 of the substance: cool the mixture if the reaction becomes too violent. Shake the mixture for two minutes and then pour it into about 10 cm^3 of cold water: the nitro compound separates as a dense yellow oil or solid. Branched chain alkanes are more readily nitrated than straight chain alkanes.

8 To 1 cm^3 of the substance add slowly 1 cm^3 of concentrated sulphuric acid and shake carefully: a homogeneous solution should be obtained. Pour the mixture into about 10 cm^3 of cold water, stirring all the time: a clear solution of the sulphonic acid may be obtained.

Record all observations made and write equations for reactions that occur.

4.5 The Photochemical Bromination of an Alkane

Take two test-tubes and wrap one thoroughly in black paper so that no light can enter it. Clamp the two tubes side by side and put 5 cm^3 of cyclohexane in each. Set up an ultra-violet light to shine horizontally on the cyclohexane at a range of about 5 cm; fix a teat pipette just above each test-tube, then put 1 cm^3 of a 5% solution of bromine in carbon tetrachloride (tetrachloromethane) in each pipette.

Add the bromine solution slowly to the alkane.
a) Observe what happens and in each case test for hydrogen bromide with moist blue litmus paper and a glass rod dipped in a concentrated solution of ammonia.
b) Add 3 cm^3 of water to each of the test-tubes. Separate the aqueous and organic layers in a funnel. To each layer of each tube add 1 cm^3 of a solution of silver nitrate in ethanol and shake well.

4.6 The Cracking of an Alkane

1 The constituents of crude oil having a relatively high molecular mass are in the majority but they are not the most useful part, so the situation is remedied by 'cracking', i.e. thermal or catalytic decomposition.

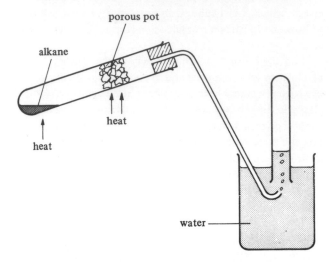

Set up the apparatus as shown and put $2\,cm^3$ of the alkane ('medicinal paraffin') carefully into the bottom of the boiling tube (150 x 25 mm). Some anti-bumping granules should be put in the alkane. Heat the tube at the points indicated in the figure concentrating the heat on the porous pot and collect several test-tubes of the gas produced. Cork the test-tubes and remove the delivery tube from the beaker as soon as heating ceases.

2 Test the gas produced as follows.
   a) Add a mixture of $1\,cm^3$ of potassium permanganate [manganate(VII)] solution and $1\,cm^3$ of sodium carbonate solution (Baeyer's reagent), recork the test-tube and shake well.
   b) Add $1\,cm^3$ of bromine water, recork the test-tube and shake well.
   c) To the last tube of gas collected apply a lighted splint (all the air should have been displaced from the apparatus before this was collected).

3 If glass beads are used in (1) instead of porous pot are the same results obtained?

4 The experiment may be repeated with mineral oil and steel wool.

4.7 The Preparation and Properties of Ethylene (Ethene)

Alternative preparations are given in (1), (2) and (3).

1 Set up the apparatus as in experiment 4.6 but use ethanol with anti-bumping granules in the tube and concentrate the heating on the porous pot. Collect several test-tubes of the ethylene (ethene) produced and remove the delivery tube from the beaker as soon as heating ceases.

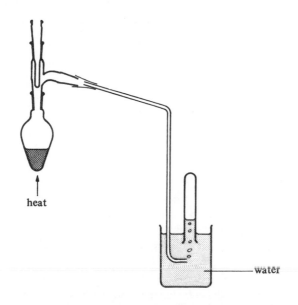

heat

water

2 Set up the apparatus as in the figure. Into the flask put 10 cm³ of ethanol and a few anti-bumping granules. Then slowly add with constant swirling 16 cm³ of concentrated (i.e. syrupy, 85%, 16M) phosphoric acid. Heat the flask. Collect several test-tubes of ethylene and remove the delivery tube from the beaker as soon as heating ceases.

3 A third alternative is to boil ethanol in a flask or a boiling tube, to pass the vapour over aluminium oxide in a combustion tube and to collect the ethylene produced over water as before.

4 Repeat the tests suggested in experiment 4.6 (2).

4.8 The Preparation of Cyclohexene

1 Put 5 cm³ of concentrated phosphoric acid (syrupy, 85%, 16M) in the flask together with a few anti-bumping granules and set up the apparatus in a fume cupboard as in figure 4.F. The flask should be heated in an oil bath (medicinal paraffin or MS 550 silicone oil) to 440K (167°C). The receiver should be cooled in ice; water should not run in the vertical condenser. Over a period of an hour add 22 cm³ of cyclohexanol.

Replace the teat pipette by a thermometer in its pocket and raise the temperature of the bath to 470 K (197°C), but do not allow the temperature of the top of the column to exceed 360K (87°C).

2 The distillate obtained as in (1) is cyclohexene together with a little water. Saturate the distillate using fine crystals of sodium chloride and separate the upper layer of cyclohexene. Dry the cyclohexene with anhydrous magnesium sulphate.

3 Set up the apparatus for distillation (figure 4.C) using an air bath (an empty tin surrounding the flask) for heating purposes. Decant the dried cyclohexene into the flask and put in a few anti-bumping granules. Heat the crude cyclohexene, using the air bath, and collect the portion boiling at 356±2K (83±2°C).

4.9 The Properties of Alkenes

2-methylbut-2-ene or cyclohexene can be studied.

1 Shake 2 cm³ of water with a few drops of the alkene. Test the water (a solution?) with pH paper.

2 Put a few drops of the substance on a crucible lid and ignite it with a lighted splint.

3 Put 1 cm³ of potassium permanganate [manganate(VII)] solution and 1 cm³ of dilute sulphuric acid in a test-tube; add a few drops of the alkene and shake well.

4 Put 1 cm³ of potassium permanganate solution and 1 cm³ of sodium carbonate solution (Baeyer's reagent) in a test-tube; add a few drops of the alkene and shake well.

5 Put 1 cm³ of bromine water in a test-tube; add a few drops of the alkene and shake well.

6 Put 0.5 cm³ bromine dissolved in carbon tetrachloride (tetrachloromethane) in a test-tube, add a few drops of the alkene, observe, then wash away down the fume cupboard drain.

7 To 1 cm³ of the substance add slowly 1 cm³ of concentrated sulphuric acid, shake carefully and note any change in temperature or colour. Dilute the solution with 3 cm³ of water and shake again; if nothing is apparent add several spatula loads of sodium chloride to the solution.

Record all observations made and write equations for reactions that occur. Compare the results obtained with those obtained using alkanes.

4.10 The Reversible Polymerization of Methyl Methacrylate (Methyl 2-methylpropenoate)

Care: this experiment must be done in a fume cupboard.

1 As the polymer itself is probably the most readily available material the experiment starts with it: the polymer of methyl methacrylate, $CH_2 = C(CH_3)COOCH_3$, a derivative of ethylene (ethene), is Perspex.

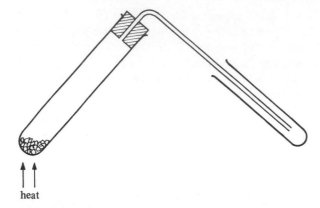

heat

Heat about four spatula loads of small pieces of Perspex in a test-tube, as shown in the figure, collecting the distillate, which is the monomer, in the cold test-tube.

2 To the liquid in the test-tube add a few crystals of benzoyl peroxide [di(benzoyl) peroxide] which is kept moist because of its explosive nature. Cork the test-tube and shake it to mix the contents. Loosen the cork and place the test-tube in a beaker of cold water. At a moderate speed bring the water to the boil and leave it boiling for up to 30 minutes: observe the effect on the liquid monomer.

4.11 The Properties of Aromatic Hydrocarbons

Benzene (vapour **poisonous**) and toluene (methylbenzene) can be referred to as arenes.

1 Shake $2 \, cm^3$ of water with a few drops of the arene. Test the water (a solution?) with pH paper.

2 Put a few drops of the substance on a crucible lid and ignite it with a lighted splint.

3 Put $1 \, cm^3$ of potassium permanganate [manganate(VII] solution and $1 \, cm^3$ of dilute sulphuric acid in a test-tube; add a few drops of the substance and shake well.

4 Put $1 \, cm^3$ of potassium permanganate solution and $1 \, cm^3$ of sodium carbonate solution (Baeyer's reagent) in a test-tube; add a few drops of the substance and shake well.

5 Put $1 \, cm^3$ of bromine water in a test-tube; add a few drops of the substance and shake well.

6 Put $0.5 \, cm^3$ of bromine dissolved in carbon tetrachloride (tetrachloromethane) in a test-tube, add a few drops of the substance, observe before and after adding a few iron filings, then wash away down the fume cupboard drain.

7 To $1 \, cm^3$ concentrated nitric acid add $1 \, cm^3$ concentrated (or fuming) sulphuric acid and then carefully add $0.5 \, cm^3$ of the substance: cool the mixture if the reaction becomes too violent. Shake the mixture for two minutes and then pour it into about $10 \, cm^3$ of cold water: the nitro compound separates as a dense yellow oil or solid.

8 To $1 \, cm^3$ of the substance add slowly $1 \, cm^3$ of concentrated sulphuric acid and shake carefully: a homogeneous solution should be obtained. Pour the mixture into about $10 \, cm^3$ of cold water, stirring all the time: a clear solution of the sulphonic acid is obtained.

Record all observations made and write equations for reactions that occur. Compare the results obtained with those obtained using alkenes and alkanes.

4.12 The Preparation and Properties of Ethyne—Acetylene

This is done as a demonstration because of the explosive nature of ethyne-air mixtures.

1 Ethyne is prepared and collected (either in small gas-jars or boiling tubes) as suggested in the figure. For tests (f) and (g) pass the gas into a boiling tube containing the reagent.

2 Test the ethyne, the first member of the homologous series of alkynes, as follows, recorking the test-tube in experiments (a) to (d) in order to be able to shake it well.

a) Put 1 cm³ of potassium permanganate [manganate (VII)] solution and 1 cm³ of dilute sulphuric acid in a test-tube; add the solution to the ethyne and shake well.

b) Put 1 cm³ of potassium permanganate solution and 1 cm³ of sodium carbonate solution (Baeyer's reagent) in a test-tube; add the solution to the ethyne and shake well.

c) Put 1 cm³ of bromine water in a test-tube; add the solution to the ethyne and shake well.

d) Put 0.5 cm³ of bromine dissolved in carbon tetrachloride (tetrachloromethane) in a test-tube; add the solution to the ethyne, shake well, observe then wash away the liquid down the fume cupboard drain.

e) Put a lighted splint to a gas-jar (or boiling tube) full of ethyne behind a safety screen.

f) Pass ethyne into a boiling tube containing copper(I) chloride dissolved in ammonia solution. See experiment 2.25. Destroy the copper acetylide [copper(I) dicarbide] by pouring it into 10 cm³ of concentrated hydrochloric acid.

g) Add a few drops of sodium hydroxide solution to 2 cm³ of silver nitrate solution, causing the precipitation of silver oxide, then add the minimum quantity of dilute ammonia solution to dissolve the precipitate. Pass ethyne into this ammoniacal solution of silver nitrate. Filter off the precipitate and hold the screwed up filter-paper by a wire about 10 cm above a Bunsen burner flame.

Compare the results obtained with those obtained using alkanes, alkenes and aromatic hydrocarbons (arenes).

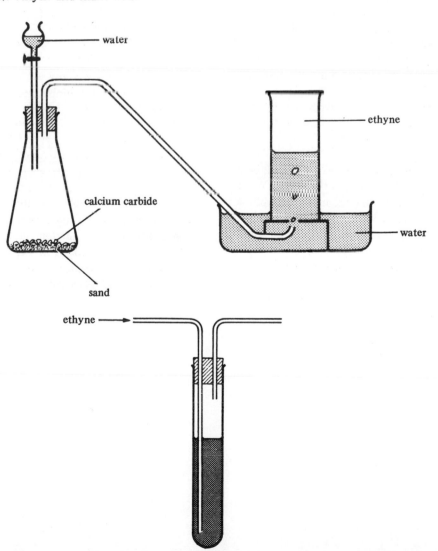

Halogen Derivatives of Hydrocarbons

4.13 The Preparation of Bromoethane — Ethyl Bromide

This is to illustrate the process of replacement of an —OH by a —Br group.

1 Put 6 cm^3 of ethanol (4.8g) in the flask and carefully add 5 cm^3 of concentrated sulphuric acid with shaking and cooling. Then add 12g of powdered potassium bromide. These quantities represent about 0.1 mole of each reagent.

   Set up the flask for distillation (figure 4.D); have the delivery tube just projecting under the surface of some water in the receiver. Attach a piece of rubber tubing to the other tube in the receiver and put its free end in the drain to carry away any ethene vapour.

   Heat the flask slowly and carefully to avoid undue frothing of the contents. Bromoethane slowly distils at about 312K (39°C) and falls to the bottom of the receiver as drops of oil: continue until no more is seen to distil.

2 Transfer the contents of the receiver to a separating funnel. Run off the lower layer of bromoethane into a test-tube and discard the water. Wash the bromoethane by shaking it with two volumes of sodium carbonate solution; the aqueous layer can be removed with a teat pipette. What does this do? Finally wash the bromoethane with an equal volume of water; remove the water as before. Then dry the bromoethane by adding a few pieces of anhydrous calcium chloride and shaking the flask gently at five minute intervals for 20 minutes.

3 Decant the dried bromoethane, which should be a clear colourless liquid, back into the apparatus set up for distillation. This time the receiver should be a small dry flask preferably cooled in ice. Add a few anti-bumping granules and carry out the distillation: collect the portion distilling at 312 ± 2K (39 ± 2°C). The density of bromoethane is 1.46 g/cm^3.

   Write the equations for the main and side reactions. Calculate the percentage yield.

4.14 The Preparation of 1-Bromobutane — Butyl Bromide

This experiment illustrates the process of replacement of an —OH by a —Br group.

1 Put 12g of sodium bromide, 12 cm^3 of water and, accurately, 9 cm^3 of butan-1-ol (density 0.81 g/cm^3) into the flask. The quantities represent about 0.1 mole of each of the reagents. Set up the apparatus as in figure 4.B having the flask dipping into a beaker of water. Slowly add 10 cm^3 of concentrated sulphuric acid swirling the flask so that the reagents mix well.

2 Heat the flask, without a water bath, so that the reaction mixture just boils for 30–45 minutes.

3 Allow the apparatus to cool for a few minutes and then rearrange it for distillation (figure 4.D). Heat the mixture and collect about 12 cm^3 of distillate in a test-tube: this is a form of steam distillation and at this stage no more oily drops should be seen coming off with the water. Dismantle the apparatus and dry it.

4 Separate the bromobutane from the water using a separating funnel or teat pipette. Rinse the test-tube with two 5 cm^3 portions of concentrated hydrochloric acid and add these portions to the bromobutane in the funnel. Any unreacted butanol is removed by the acid. Next wash the bromobutane with an equal volume of 1M sodium hydrogencarbonate solution. Finally dry the bromobutane with small portions of anhydrous sodium sulphate.

5 Set up the dried apparatus for distillation (figure 4.D). Pour the dried bromobutane through a micro Hirsch funnel containing a filter-paper to retain the desiccant. Distil the bromobutane collecting the fraction that boils at 375 ± 2K (102 ± 2°C).

   Measure the volume of product (its density is 1.28 g/cm^3) and hence calculate the percentage yield.

108

4.15 The Preparation of Iodoethane — Ethyl Iodide

This is to illustrate the substitution of an —OH by an —I group.

---

1 Set up the apparatus for refluxing (figure 4.A) and put 0.5 g of red phosphorus and 5 cm^3 of ethanol (4 g) directly into the flask. Weigh out 5 g of iodine and divide it into five portions: add these at intervals directly into the flask, replacing the reflux condenser and cooling the flask between each addition. Leave the mixture to stand, with occasional shaking, for at least an hour; the condenser water does not need to be run during this period.

---

2 With the condenser now running, heat the mixture on a water bath for two hours. Then rearrange the apparatus for distillation (figure 4.C), keeping the water bath, and distil, collecting all vapours having a boiling-point below 353 K (80°C).

---

3 Transfer the contents of the receiver to a separating funnel. Run off the lower layer of iodoethane into a test-tube and discard the water. Wash the iodoethane by shaking it with twice its volume of sodium carbonate solution; the aqueous layer can be removed with a teat pipette. What does this do? Finally wash the iodoethane with an equal volume of water; remove the water as before. Then dry the iodoethane by adding a few pieces of anhydrous calcium chloride and shaking the flask gently at five minute intervals for 20 minutes.

---

4 Decant the dried iodoethane back into the apparatus set up for distillation. Add a few anti-bumping granules and carry out the distillation: collect the portion distilling at 345 ± 2 K (72 ± 2°C). Iodoethane has a density of 1.94 g/cm^3. It may be discoloured due to slight decomposition: add a crystal of sodium thiosulphate [thiosulphate(VI)] to remove this discoloration.

   Write the equations for the reactions concerned. Calculate the percentage of the theoretical yield obtained.

4.16 The Preparation of 1,2-Dibromoethane— Ethylene Dibromide

This is to illustrate an addition reaction of an alkene.

1 Set up the apparatus, as in 4.7(2) but instead of collecting the ethene over water pass it through three boiling tubes as in this figure.

   The first boiling tube should contain about 40 cm³ of concentrated sodium hydroxide solution (5M) and the second and third tubes each 2 cm³ of bromine covered with water to a depth of about 5 cm. An absorption tube containing loosely packed soda-lime may be used to prevent any bromine vapour escaping. The tubes containing bromine should be cooled by immersion in a beaker of water.

   Care: an empty boiling tube should be inserted between the distillation flask and the first boiling tube illustrated. Bromine is toxic and the experiment should be done in a fume-cupboard.

Pass ethene through the bromine until the colour of the latter has almost vanished. Turn off the Bunsen burner and immediately disconnect the boiling tubes from the adaptor.

2 The bromine has been converted into 1,2-dibromoethane. Pour the contents of one boiling tube into the other and add 1 cm³ of dilute sodium hydroxide solution. Stopper and shake the tube; remove the aqueous layer with a teat pipette. Continue in this manner until the colour due to bromine has just vanished. Next wash the dibromoethane with two 5 cm³ portions of water. Dry the dibromoethane with anhydrous calcium chloride in a corked test-tube.

3 Distil the product in the usual apparatus (figure 4.C) collecting the fraction boiling at 404 ± 2K (131 ± 2°C).

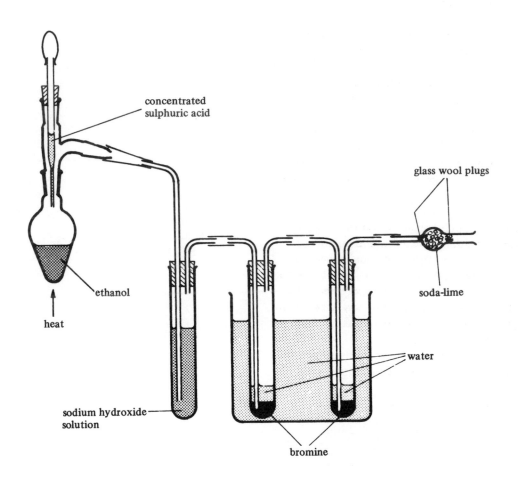

concentrated sulphuric acid

glass wool plugs

soda-lime

ethanol

heat

water

sodium hydroxide solution

bromine

4.17 The Preparation of 2-Chloropropane

This is to illustrate the substitution of an —OH by a —Cl group. The first three stages must be done in a fume cupboard.

---

1   Put 25 cm^3 of thionyl chloride (sulphur dichloride oxide, SOCl$_2$) and a few anti-bumping granules into the apparatus set up for distillation as in figure 4.D plus an absorption tube filled with anhydrous calcium chloride to prevent moisture getting into the apparatus. Distil the thionyl chloride, collecting the portion boiling at 352 ± 2 K (79 ± 2 °C) — 15 cm^3 are required. Tip the residue in the distillation flask down the fume cupboard drain together with plenty of water and allow the flask to drain for a moment.

---

2   Put 12 cm^3 of **propan-2-ol** in the distillation flask plus a few anti-bumping granules and set up the apparatus for refluxing (figure 4.A plus a water bath). Carefully add five 3 cm^3 portions of thionyl chloride, swirling the flask between each addition. Then heat the water bath and maintain it boiling gently for 30 minutes.

---

3   Take away the water bath and after a moment re-arrange the apparatus for distillation (figure 4.D); the receiver should be cooled in ice. Distil off the 2-chloropropane, collecting the portion boiling at 309 ± 2 K (36 + 2 °C)

---

4   Transfer the contents of the receiver to a separating funnel. Wash the distillate by shaking it with an equal volume of sodium carbonate solution and then with an equal volume of water. Then dry the 2-chloropropane by leaving it in contact with some granular anhydrous calcium chloride for at least 15 minutes. Redistil the 2-chloropropane as in part (3).

4.18 The Properties of Halogen Derivatives of the Aliphatic Hydrocarbons

Bromoethane, iodoethane, 1- or 2-chloropropane, chloroform (trichloromethane), and carbon tetrachloride (tetrachloromethane) are simple examples of the halogen derivatives of the aliphatic hydrocarbons. For a direct comparison the 1-halobutanes can be employed; another experiment may be to compare the four isomers of C$_4$H$_9$Br.

---

1   Add a few drops of the halide to 2 cm^3 water. Shake vigorously. Do they mix? Test the water (a solution?) with pH paper.

---

2   Add a few drops of the substance to 2 cm^3 of silver nitrate solution and shake vigorously for several minutes. Repeat the experiment having the silver nitrate solution in a test-tube in a beaker of water at about 330–340K (about 57–67°C).

---

3   Put 1 cm^3 of the substance with 10 cm^3 of sodium hydroxide solution in ethanol in a small flask fitted with a reflux condenser; heat for ten minutes. If a white solid has separated dilute the solution with water. Acidify the solution with dilute nitric acid and then add silver nitrate solution.

4.19 The Properties of Halogen Derivatives of the Aromatic Hydrocarbons

Chlorobenzene, benzyl chloride [(chloromethyl)-benzene], benzylidene dichloride or benzal chloride [(dichloromethyl)benzene] and benzotrichloride [(trichloromethyl)benzene] are simple examples of the halogen derivatives of the aromatic hydrocarbons. They are all liquids; the first substance has a typical aromatic odour but the other substances have irritating or even lachrymatory vapours.

1 Add a few drops of the halide to 2 cm³ water. Shake vigorously. Do they mix?

2 Add a few drops of the substance to 2 cm³ of silver nitrate solution and shake vigorously for several minutes. Repeat the experiment having the silver nitrate solution in a test-tube in a beaker of water at about 330–340K (57–67°C). Is it possible to distinguish between nuclear and side-chain chlorine?

3 Put 1 cm³ of the substance with 10 cm³ of sodium hydroxide solution in ethanol in a small flask fitted with a reflux condenser; heat for ten minutes. If a white solid has separated dilute the solution with water. Acidify the solution with dilute nitric acid and then add silver nitrate solution.

4 Working in a fume cupboard, put 0.5 cm³ of chlorobenzene in a test-tube and add 1 cm³ of concentrated sulphuric acid. Warm the mixture gently with continuous shaking for about five minutes. Then pour the mixture into about 20 cm³ of water: an oil separates out and it will, on scratching the test-tube with a glass rod, solidify indicating that the chlorobenzene has become converted to a substance of greater relative molecular mass.

The experiment may be repeated with benzyl chloride. The characteristic odour of bitter almonds, due to benzaldehyde, also becomes noticeable.

5 Boil a mixture of 25 cm³ of saturated potassium permanganate [manganate(VII)] solution, four spatula loads of anhydrous sodium carbonate and 0.5 cm³ of benzyl chloride under reflux (figure 4.A) for 30 minutes. Filter the hot solution using a filter-paper and a stemless filter-funnel. Acidify the filtrate with concentrated hydrochloric acid: on cooling benzoic acid crystallizes out. Some benzaldehyde may be smelled.

The benzoic acid can be collected by using a Buchner filter-funnel and flask. The suction of the water-pump causes rapid draining of water from the crystals and this may be enhanced by pressing on the crystals with a clean glass stopper.

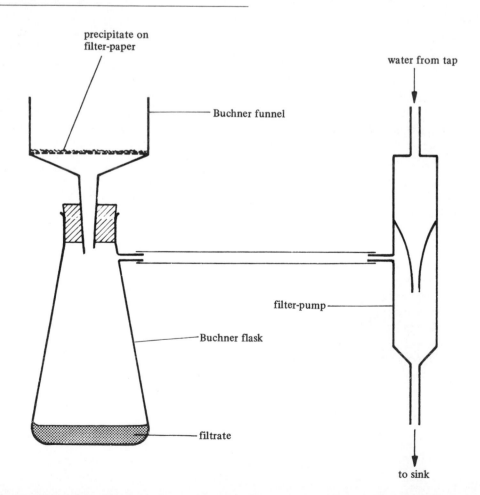

precipitate on filter-paper

Buchner funnel

water from tap

Buchner flask

filter-pump

filtrate

to sink

Hydroxy Derivatives of Hydrocarbons

4.20 The Hydration of Alkenes

1 Put 2 cm³ of water in a boiling tube and slowly add, with cooling (use an ice bath) and shaking, 6 cm³ of concentrated sulphuric acid. Put an equal volume of an alkene, e.g. cyclohexene, in another boiling tube and cool the two tubes in the ice bath.

2 Slowly add the moderately concentrated sulphuric acid to the alkene, with cooling and shaking. Continue shaking until the mixture becomes homogeneous (about five minutes) and then allow the mixture to stand in the bath for an equal length of time.

3 Pour the mixture into an equal volume of water in a separating funnel: two layers should form, one of alcohol, the lower one of dilute sulphuric acid. Discard the lower layer and wash the upper layer with an equal volume of water. Dry the alcohol with anhydrous potassium carbonate.

4 (It may be advantageous at this stage, or even at stage 3, to pool the samples obtained by individuals.) Distil the alcohol, collecting the appropriate fraction.

4.21 The Properties of Aliphatic Alcohols

Methanol, ethanol, propan-1-ol and propan-2-ol may be studied.

1 Shake 2 cm³ of water with a few drops of the alcohol. Test the water (a solution?) with pH paper.

2 Put a few drops of the alcohol on a crucible lid and set fire to it: observe the nature of the flame.

3 In a fume cupboard: to 1 cm³ of the alcohol in a dry test-tube carefully add a spatula load of phosphorus pentachloride. There is a vigorous evolution of a pungent smelling poisonous gas: identify it.

4 To 0.2 cm³ of the alcohol add 2 cm³ 0.5 M potassium iodide solution and 4 cm³ sodium hypochlorite [chlorate(I)] solution, warm the mixture to 323 K (50°C) for two minutes and then cool it. If the alcohol is ethanol or propan-2-ol fine yellow crystals of iodoform (triiodomethane) are produced; the test depends on the formation by oxidation or the initial presence of the CH_3CO- group.

5 Methanol can be identified by converting it to methyl salicylate (2-hydroxybenzoate) the odour of which (oil of wintergreen) is characteristic and easily recognized.

Heat 1 cm³ of methanol with a spatula load of salicylic acid and a few drops of concentrated sulphuric acid for one minute. Cool and pour the mixture into about 10 cm³ of water in a small beaker and note the smell of any vapour.

6 Ethanol is also confirmed by putting 1 cm³ of the substance into a test-tube then adding 1 cm³ of glacial acetic (ethanoic) acid and a few drops of concentrated sulphuric acid. Heat the tube for one minute and then pour the mixture into about 10 cm³ of water in a small beaker: the fruity odour of ethyl acetate (ethanoate) should be apparent.

7 Make up 4 cm³ of a concentrated solution of potassium dichromate [dichromate(VI)] and add 1 cm³ of concentrated sulphuric acid, warm the mixture gently and then add 0.5 cm³ of ethanol: the sweet smell of acetaldehyde (ethanal) should become apparent and it is replaced by the sharp smell of acetic acid. Explain the colour change observed.

8 **The Lucas test**
This is done to distinguish between primary, secondary and tertiary alcohols. Lucas' reagent is zinc chloride dissolved in concentrated hydrochloric acid.

Put 1 cm³ of the alcohol into a test-tube. Add 6 cm³ of Lucas' reagent, cork the test-tube and shake it. Allow it to stand for five minutes. With primary alcohols of low relative molecular mass the aqueous layer remains clear; with secondary alcohols chlorides separate on standing; with tertiary alcohols separation takes place immediately.

9 Dissolve a small piece of sodium in 2 cm³ of ethanol; if the reaction ceases before all the sodium has dissolved add a little more ethanol. Then add 0.5 cm³ of iodoethane and stand the test-tube in a beaker of warm water. No flames should be near. Diethyl ether (ethoxyethane), which is very flammable, should be smelled.

113

4.22 The Properties of Benzyl Alcohol (Phenylmethanol)

1 To $2\,cm^3$ of water add a few drops of benzyl alcohol: do they mix? Compare ethanol.

2 Put a few drops of benzyl alcohol on a crucible lid and set fire to it: observe the nature of the flame.

3 In a fume cupboard: to $1\,cm^3$ of the alcohol in a dry test-tube carefully add a spatula load of phosphorus pentachloride. Observe what happens and test the gas evolved with damp pH paper.

4 To $0.5\,cm^3$ of benzyl alcohol add $2\,cm^3$ of dilute sulphuric acid and $2\,cm^3$ of potassium permanganate [manganate(VII)] solution. Heat the mixture to boiling and then allow it to cool: white crystals of benzoic acid form.

5 Put $0.5\,cm^3$ of glacial acetic acid in a test-tube and add $1\,cm^3$ of benzyl alcohol followed by a few drops of concentrated sulphuric acid. Heat the tube carefully for one minute and then pour the mixture into about $10\,cm^3$ of water in a beaker: the jasmin odour of benzyl acetate (ethanoate) should be apparent.

4.23 The Properties of Phenol

Care: Phenol and its aqueous solution are very corrosive and are dangerous if they touch the skin.

1 To $2\,cm^3$ of water slowly add phenol crystals until no more dissolve. Observe the effect of stirring in $2\,cm^3$ of sodium hydroxide solution. Then add concentrated hydrochloric acid dropwise with stirring.

2 To $2\,cm^3$ of an aqueous solution of phenol add half a spatula load of sodium hydrogencarbonate.

3 To $2\,cm^3$ of an aqueous solution of phenol add half a spatula load of anhydrous sodium carbonate.

4 Make up some neutral iron(III) chloride solution: to $1\,cm^3$ iron(III) chloride solution add sodium hydroxide solution until a faint precipitate appears, centrifuge and add a drop of the solution to a very dilute solution of phenol. Note the coloration.

5 To a crystal of phenol dissolved in $2\,cm^3$ of diethyl ether (ethoxyethane) add a very small piece of sodium. (To destroy any excess of sodium safely add $3\,cm^3$ of ethanol and then pour the mixture into $20\,cm^3$ of water before pouring away down the drain.) **Care: diethyl ether is very flammable.**

6 To $2\,cm^3$ of an aqueous solution of phenol slowly add bromine water.

7 **Liebermann's reaction**
Put a few crystals of sodium nitrite [nitrate(III)] and one spatula load of phenol in a clean dry test-tube and then heat for 20 seconds. Cool and add twice the volume of concentrated sulphuric acid, stir and pause 1–2 minutes. Note the initial colour, then add water carefully and finally add sodium hydroxide solution.

8 Put one spatula load of phenol crystals and one spatula load of phthalic (benzene-1,2-dicarboxylic) anhydride in a dry test-tube. Add two drops of concentrated sulphuric acid and warm gently for one minute with great care. Cool and add sodium hydroxide solution in excess. The indicator phenolphthalein has been produced.

9 **Schotten-Baumann reaction**: see experiment 4.47.

Ethers

4.24 The Preparation of Diethyl Ether (Ethoxyethane)

This process is an example of nucleophilic substitution (of ethyl hydrogensulphate by ethanol) although it is often referred to as a partial dehydration of an alcohol. It should be done as a demonstration because of the flammable nature of the product.

1 Put $10\,cm^3$ of ethanol in the reaction flask and slowly add $8\,cm^3$ of concentrated sulphuric acid, with constant swirling and cooling under the tap.

2 Set up the apparatus as in figure 4.F but omitting the vertical condenser. Attach a piece of rubber tubing to the receiver and lead it well away from the experimental area. Cool the receiver with ice. A tin can with a few holes punched in the lid serves as an air bath so that the flask can be heated to about $425\,K$ (about $150\,°C$) — alternatively an oil bath, e.g. medicinal paraffin, may be employed. Heat the flask and when the ether begins to distil over drip in another $10\,cm^3$ of ethanol.

3 Turn off the Bunsen burner and put the distillate in a separating funnel. Wash the ether twice with an equal volume of sodium hydroxide solution. Diethyl ether has a density of $0.71\,g/cm^3$. To remove any ethanol which has distilled wash the ether with an equal volume of a saturated solution of calcium chloride in water. Dry the ether in a test-tube with anhydrous calcium chloride (any remaining ethanol will also be absorbed).

4 Set up the apparatus as in figure 4.D and use a beaker of warm water ($330\,K$, about $60\,°C$) to heat the flask. Collect the fraction which distils at $308 \pm 2\,K$ ($35 \pm 2\,°C$); again cool the receiver with ice.

4.25 The Preparation of Butyl Ethyl Ether (Ethoxybutane)

1 Put $1\,g$ of clean sodium in the form of finely divided pieces into the flask and set up the apparatus as in figure 4.A. Add $20\,cm^3$ of ethanol down the condenser and allow the sodium to dissolve completely before proceeding: should the reaction become very vigorous cool the flask. Next add $3\,cm^3$ of 1-bromobutane down the condenser and reflux the mixture on a water bath for at least 30 minutes.

2 Add $15\,cm^3$ of water to dissolve the salt that has separated during refluxing. Transfer the solutions to a separating funnel and discard the lower aqueous layer. Add $15\,cm^3$ of water to wash the ether. Finally dry the ether with anhydrous calcium chloride.

3 Filter the dried ether into the flask and set up the apparatus as in figure 4.D. Distil off the ether, collecting the portion boiling between $365 \pm 2\,K$ ($92 \pm 2\,°C$).

115

4.26 The Preparation of Acetaldehyde (Ethanal)

This experiment illustrates the partial oxidation of a primary alcohol.

1 Put 8 cm³ of water into the flask and carefully add 3 cm³ of concentrated sulphuric acid, with swirling and cooling. Set up the apparatus as in figure 4.F but without the vertical condenser. Put 8 g of sodium dichromate (**care**: corrosive) in a small beaker and add 8 cm³ of water and 6 cm³ of ethanol; stir until a homogeneous solution is obtained.

2 Heat the flask until the contents are boiling gently, then reduce the flame. Add the solution from the beaker in 2 cm³ portions so that the solution in the flask again boils gently. Continue heating the flask until about 10 cm³ of distillate have been obtained: this is an aqueous solution of acetaldehyde.

What do you see happening during the reaction? Write the equation for the reaction making use of the partial equation

$$C_2H_5OH \rightarrow CH_3CHO + 2H^+ + 2e^-$$

This method may also be tried using propanol in place of ethanol; the boiling-point of propionaldehyde (propanal) is 321K (48°C).

4.27 The Properties of Aliphatic Aldehydes

Formaldehyde (methanal; a 40% m/V solution in water is called formalin), acetaldehyde (ethanal) or butyraldehyde (butanal) may be studied.

1 Shake 2 cm³ of water with a few drops of the aldehyde. Test the water (a solution?) with pH paper.

2 Put a few drops of the aldehyde on a crucible lid and ignite it with a lighted splint.

3 To 1 cm³ of the substance add 1 cm³ of sodium carbonate and then 2 cm³ of Fehling's solution (1 cm³ of A and of B if supplied separately). Boil gently for one minute.

4 Put 3 cm³ of silver nitrate solution into a thoroughly clean test-tube, add three drops of sodium hydroxide solution, then ammonia solution dropwise until the precipitate has just redissolved (Tollen's reagent), and finally add three drops of the substance and gently warm, with shaking.

5 To 0.5 cm³ of the substance add 3 cm³ of potassium dichromate [dichromate(VI)] solution and a few drops of concentrated sulphuric acid. Warm the mixture gently, then smell cautiously.

6 To 0.5 cm³ of the substance add 3 cm³ of potassium permanganate [manganate(VII)] solution and a few drops of concentrated sulphuric acid. Warm the mixture gently, then smell cautiously.

7 To 0.2 cm³ of the substance add 2 cm³ of 0.5M potassium iodide solution and 4 cm³ of sodium hypochlorite [chlorate(I)] solution. Warm the mixture to 323K (50°C) for two minutes and then cool it. (Alternative reagents: 0.2 cm³ of the substance, 1 cm³ of a solution of iodine in aqueous potassium iodide and add sufficient sodium hydroxide solution just to remove the colour).

8 To 0.5 cm³ of the cold substance add 1 cm³ of Schiff's reagent.

9 To 0.5 cm³ of the substance add 3 cm³ of saturated sodium hydrogensulphite [hydrogensulphate(IV)] solution. Shake the mixture and then let it stand.

10 To 0.5 cm³ of the substance add 2 cm³ of 2,4-dinitrophenylhydrazine solution (in 3M hydrochloric acid, Brady's reagent). If shaking has produced no apparent reaction, warm then cool the mixture.

11 Warm 1 cm³ of the substance with 1 cm³ of concentrated sodium hydroxide solution, then smell cautiously.

12 To 1 cm³ of the substance add one drop of concentrated sulphuric acid. Pause for a minute then cool the mixture and pour it into 10 cm³ of water in a beaker: observe its appearance.

4.28 The Properties of Aliphatic Ketones

Acetone (propanone), ethyl methyl ketone (butanone) and cyclohexanone are suitable for study.

Repeat experiment 4.27 using a ketone. Compare and contrast your results with those obtained using an aldehyde.

4.29 The Properties of Benzaldehyde

Note the odour and then repeat experiment 4.27 using benzaldehyde. Compare and contrast your results with those obtained using aliphatic aldehydes.

In part (11) stir the mixture while it is heated for about five minutes. Add 5 cm³ of water, stir again and then decant off the aqueous solution of sodium benzoate. The reaction is one of disproportionation and was first studied by Cannizzaro. To the sodium benzoate solution add 1 cm³ of concentrated hydrochloric acid and, if necessary, cool the mixture. The impure organic product left after the dissolution of the sodium benzoate is benzyl alcohol (phenylmethanol).

4.30 The Preparation of Acetophenone (Phenylethanone, Methyl Phenyl Ketone)

Friedel-Craft's method is used to illustrate the process of acetylation.

1 Dry about 30 cm³ of benzene with some small pellets of sodium and distil 20 cm³ of acetyl (ethanoyl) chloride in a dry apparatus set up as in figure 4.D with a loosely packed anhydrous calcium chloride tube preventing the entry of moisture.

2 Set up the apparatus as in figure 4.B and attach, by a piece of rubber tubing, an inverted funnel to the side of the adaptor. The lip of the funnel should just dip under the surface of some water in a beaker. Into the flask put 25 cm³ of the dry benzene and 10 g of powdered anhydrous aluminium chloride and cool the flask in water. Add the acetyl chloride in four 2 cm³ portions at three minute intervals being careful to mix the substances thoroughly after each addition.

3 Heat the flask to 320 K (about 50°C) for at least 40 minutes: the evolution of hydrogen chloride should have diminished considerably by then, indicating the completion of the reaction.

4 Allow the mixture to cool and then pour it, with stirring, into about 75 cm³ of ice and water. If any solid separates add a little concentrated hydrochloric acid. Using a separating funnel separate the oily layer of acetophenone and benzene. Wash the organic layer with an equal volume of sodium hydroxide solution and then with water. Dry the organic layer with anhydrous calcium chloride.

5 Distil the acetophenone and benzene in the apparatus set up as in figure 4.D: collect the fraction boiling at 475 ± 2 K (202 ± 2°C). The water should be drained from the condenser when the temperature of the vapours inside reaches 425 K (152°C).

4.31 The Preparation of Benzophenone (Diphenylmethanone, Diphenyl Ketone)

Modify the previous experiment 4.30 for acetophenone (phenylethanone) to prepare benzophenone. A fume cupboard must be used when benzoyl chloride is employed because it is very lachrymatory.

4.32 The Preparation of Derivatives for Melting-Point Determinations

Crush the substance with the sealed end of a melting-point tube on a filter-paper or with a pestle in a clean mortar and place the powdered solid to a depth of 1 cm in a capillary (melting-point) tube closed at one end. The tube is then placed alongside a thermometer in a beaker of silicone oil (e.g. MS 550) and the beaker then slowly heated. Surface tension holds the tube to the thermometer. The rate of heating is decreased as the expected melting-point is approached: the crystals then collapse to a clear liquid.

A sharp melting-point indicates a pure substance (or a eutectic mixture) and gradual fusion indicates an impure substance or a substance undergoing decomposition. Two or three experiments may be necessary to obtain an accurate answer, but a tube can only be used once.

A liquid substance must be converted into a solid derivative. The examples chosen are for aldehydes and ketones.

1 The 2,4-dinitrophenylhydrazone

To 1 cm^3 of the substance add 5 cm^3 of 2,4-dinitrophenylhydrazine (in 3M hydrochloric acid, Brady's reagent). Cork the flask and shake the mixture thoroughly: yellow crystals of the 2,4-dinitrophenylhydrazone soon separate out. The separation should be allowed to continue for about 15 minutes with occasional shaking before the crystals are filtered off and rinsed, first with very dilute acetic acid and then with water. The hydrazone should be recrystallized from ethanol. The crystals should be filtered off and dried in a desiccator before their melting-point is determined.

Derivative of	T/K	T/°C
Formaldehyde (methanal)	439	166
Acetaldehyde (ethanal) [two forms]	420	147
or less likely	441	168
Propionaldehyde (propanal)	428	155
Butyraldehyde (butanal)	399	126
Benzaldehyde	510	237
Acetone (propanone)	399	126
Ethyl methyl ketone (butanone)	388	115
Acetophenone (phenylethanone)	523	250
Benzophenone (diphenylmethanone)	512	239
4 – Methylbenzophenone	472	199

2 The oxime

Dissolve 3.5g of hydroxyammonium chloride (NH$_3$OH$^+$ Cl$^-$, hydroxylamine hydrochloride) in 5 cm^3 of water in a small flask; dissolve 4 g of sodium acetate in 10 cm^3 of water in a boiling tube. Cool the solutions to 280K (7°C) and add the alkali to the salt. Cool the solutions again and slowly add 6 cm^3 of the substance with gentle shaking ensuring that the temperature does not rise above 285K (12°C). Allow the cold mixture to stand for about 15 minutes before filtering off the crystals of the oxime, contaminated by sodium chloride. The oxime should be recrystallized from the minimum possible quantity of petroleum ether [333–353 K (60–80°C) fraction], having a reflux condenser on the flask containing the crystals. Turn off the Bunsen burner and filter the hot solution to eliminate the sodium chloride. On cooling, crystals of the oxime separate and should be filtered off and dried in a desiccator before their melting-point is determined.

Derivative of	T/K	T/°C
Acetone (propanone)	319	46
Acetophenone (phenylethanone)	332	59
Benzophenone (diphenylmethanone)	417	144
4 – Methylbenzophenone	427	154

3 The semicarbazone

Put three spatula loads of powdered semicarbazinium chloride (NH$_2$CONHNH$_3^+$ Cl$^-$) and an equal quantity of anhydrous sodium acetate into 5 cm^3 of water and warm the mixture gently until a clear solution is obtained.

Add to the solution prepared above, 1 cm^3 of the substance under test dissolved in 5 cm^3 of ethanol: warm the reagents together on a water bath for 15 minutes at about 330K (57°C).

Some crystals of the semicarbazone may have already settled out and more will do so on cooling. Filter off the crystals and wash them thoroughly with water. Dry the crystals in a desiccator before determining their melting-point.

Derivative of	T/K	T/°C
Formaldehyde (methanal)	442(dec.)	169
Acetaldehyde (ethanal)	435	162
Propionaldehyde (propanal)	362 or 427	89 or 154
Butyraldehyde (butanal)	368 or 379	95 or 106
Benzaldehyde	495	222
Acetone (propanone)	463	190
Ethyl methyl ketone (butanone)	419	146
Acetophenone (phenylethanone)	471	198
Benzophenone (diphenylmethanone)	437	164
4 – Methylbenzophenone	394	121

Carboxylic Acids and their Derivatives

4.33 The Preparation of Acetic (Ethanoic) Acid

This experiment illustrates the oxidation of an alcohol without destroying the bonds between carbon atoms.

1 Put 6 cm³ of water into the flask and carefully add 4 cm³ of concentrated sulphuric acid to it with swirling and cooling. Next add 6 g of sodium dichromate (**care**: corrosive).

2 Set up the apparatus as in figure 4.A. Mix 2 cm³ of ethanol with 6 cm³ of water in a small beaker and add it in four equal portions down the condenser: shake the flask and cool it after each addition. Then heat the flask on a water bath for 20 minutes. Observe the colour change.

3 Allow the flask to cool for five minutes before re-arranging it as in figure 4.C. Heat the flask over a gauze and collect about 10 cm³ of distillate: this is mainly acetic acid (characteristic odour) but it may contain some acetaldehyde (ethanal).

    Write the equation for the reaction making use of the partial equation
    $$C_2H_5OH + H_2O \rightarrow CH_3COOH + 4H^+ + 4e^-$$
    Account for the colour change observed.

4.34 The Properties of Aliphatic Carboxylic Acids

Typical aliphatic carboxylic acids are formic (methanoic) acetic (ethanoic) and oxalic (ethanedioic) acids.

1 Shake 2 cm³ of water with a few drops (crystals) of the acid. Test the water (a solution?) with pH paper.

2 Put a few drops of the acid on a crucible lid and ignite it with a lighted splint. Compare the pH of acetic acid and hydrochloric acid (both 1 M).

3 To 2 cm³ of acetic acid add a spatula load of anhydrous sodium carbonate. Is carbon dioxide evolved?

4 Decarboxylation of an acid is usually carried out using the sodium salt: warm a spatula load of anhydrous sodium acetate mixed with two spatula loads of soda-lime (taken to be sodium hydroxide). Is the gas evolved flammable?

5 To 2 cm³ of ethanol add 0.5 cm³ of glacial acetic acid (**care**: corrosive) and then a few drops of concentrated sulphuric acid. Warm the mixture carefully and pour it into about 10 cm³ of water in a small beaker: note the odour of the ester formed.

6 To 0.5 cm³ of acetic acid add 2 cm³ of 2,4-dinitrophenylhydrazine solution (in 3M hydrochloric acid, Brady's reagent). Is there any evidence of a condensation reaction typical of a carbonyl group?

7 In a fume cupboard: to a few drops of glacial acetic acid in a dry test-tube carefully add a spatula load of phosphorus pentachloride.

8 To some iron(III) chloride solution add some sodium hydroxide solution until a faint precipitate appears. Centrifuge and use the solution. To the acetic acid add ammonia solution until the smell of the latter is apparent, then in the presence of an anti-bumping granule boil until the smell has gone, cool. Add a few drops of the neutral iron(III) chloride solution to the solution of the acid.

9 Formic acid has several properties not exhibited by acetic acid and the higher members of the homologous series:
a) To 1 cm³ of formic acid carefully add 1 cm³ of concentrated sulphuric acid and gently warm the mixture. Identify the poisonous gas evolved.
b) To 2 cm³ of dilute sulphuric acid add 1 cm³ of formic acid and then 1 cm³ of potassium permanganate [manganate(VII)] solution. Warm the mixture.
c) Put 3 cm³ of silver nitrate solution into a thoroughly clean test-tube, add three drops of sodium hydroxide solution, then ammonia solution dropwise until the precipitate has just redissolved (Tollen's reagent), and finally add three drops of formic acid and warm gently, with shaking.

10 Oxalic acid is a poisonous dicarboxylic acid: repeat tests (1) to (8), (9a) and (9b).

11 The tests can be extended to succinic (butanedioic) acid; tartaric (2,3-dihydroxybutanedioic) acid; citric (2-hydroxypropane-1,2,3-tricarboxylic) acid; and salicylic (2-hydroxybenzoic) acid. See experiment 5.18.

4.35 The Preparation of Benzoic Acid from an Ester

The experiment illustrates the process of alkaline hydrolysis (saponification) of an ester.

1 Put 2 cm³ of methyl benzoate, 10 cm³ of concentrated sodium hydroxide solution, 8 cm³ of water, 2 cm³ of methanol and a few anti-bumping granules into the flask and set up the apparatus as in figure 4.A. Heat the flask over a gauze for 20 minutes.

2 Allow the flask to cool for five minutes and then add dilute sulphuric acid until no more precipitation of benzoic acid occurs.

3 Transfer the precipitate to a funnel containing a filter-paper: wash the precipitate with water. The impure acid can be recrystallized from boiling water: dissolve it in the minimum quantity of boiling water, filter the hot solution using a stemless filter-funnel and then allow it to cool.

4.36 The Preparation of Benzoic Acid from Benzyl Alcohol (Phenylmethanol)

The experiment illustrates the oxidation of an alcohol without the destruction of carbon–carbon bonds.

1 Put 10 cm³ of water, 10 cm³ of concentrated sodium hydroxide solution, two spatula loads of potassium permanganate [manganate(VII)] crystals and a few anti-bumping granules into the flask. Shake the flask thoroughly to mix the contents. Set up the apparatus as in figure 4.B. Add 1 cm³ of benzyl alcohol dropwise down the condenser. Put the flask in a water-bath and heat until the water boils: maintain it so for about 30 minutes.

2 Allow the flask to cool for five minutes and then filter the solution away from any suspension using a Hirsch funnel. Add concentrated hydrochloric acid to the filtrate until no more precipitation of benzoic acid occurs.

3 Transfer the precipitate to a funnel containing a filter-paper; wash the precipitate with water. The impure acid can be recrystallized from boiling water: dissolve it in the minimum quantity of boiling water, filter the hot solution using a stemless filter-funnel and then allow it to cool.

4.37 The Preparation of Benzoic Acid from Toluene (Methylbenzene)

1 The first stage is to convert toluene into benzyl chloride [(chloromethyl) benzene]. Set up the apparatus as in figure 4.E in a fume cupboard. Into the flask put 20 cm³ of toluene (density 0.87 g/cm³) and a few anti-bumping granules. Calculate the mass of chlorine necessary to convert the toluene into benzyl chloride. Set up a chlorine generator (e.g. cold concentrated hydrochloric acid with potassium permanganate [manganate(VII)] crystals) and pass in chlorine until the required increase in mass has occurred. If an inverted funnel is fitted at the end of the delivery tube so that the tip is just submerged in a beaker of 3M sodium hydroxide solution, no chlorine should escape. It is necessary to do this experiment in daylight to promote radical formation.

Rearrange the apparatus for distillation (figure 4.C) and distil off the benzyl chloride, collecting the portion distilling at 452 ± 2K (179 ± 2°C).

2 Boil a mixture of 25 cm³ of saturated potassium permanganate solution, four spatula loads of anhydrous sodium carbonate and 11 cm³ of the benzyl chloride under reflux (figure 4.A) for 30 minutes. Filter the hot solution using a filter-paper and a stemless filter-funnel. Acidify the filtrate with concentrated hydrochloric acid: on cooling benzoic acid crystallizes out. The benzoic acid can be collected using a Buchner filter-funnel and flask (see experiment 4.19).

4.38 The Properties of Benzoic Acid

1 Shake 4 cm³ of water with a few crystals of benzoic acid (benzenecarboxylic acid) and warm the mixture; cool the solution obtained. Test the solution with pH paper.

2 Put a few crystals of the acid on a crucible lid and ignite it with a lighted splint.

3 To 4 cm³ of sodium hydroxide solution add a few crystals of benzoic acid. Is the acid more soluble in alkali than it is in water? Repeat using sodium carbonate solution.

4 Mix one spatula load of benzoic acid with two of soda-lime. Heat the mixture as shown in figure 4.11. **Test the flammability of the condensate.**

5 To 2 cm³ of ethanol add a spatula load of the acid and then a few drops of concentrated sulphuric acid. Warm the mixture carefully and pour it into about 10 cm³ of water in a small beaker: note the odour of the ester formed.

6 In a fume cupboard: to half a spatula load of benzoic acid in a dry test-tube, carefully add a spatula load of phosphorus pentachloride. The acid chloride produced (benzoyl chloride) has a dangerous vapour: it is lachrymatory. To the benzoyl chloride add a few drops of a concentrated solution of ammonia.

7 To 3 cm³ of potassium permanganate [manganate (VII)] solution and 1 cm³ of dilute sulphuric acid add a few grains of benzoic acid; gently warm the mixture. Is benzoic acid readily oxidized?

4.39 The Preparation of Acetyl (Ethanoyl) Chloride

This illustrates the replacement of an —OH group in an acid by a —Cl group.

1 The experiment must be performed in a fume cupboard and only if the apparatus is dry may it be successful.

2 Carefully put 6 cm³ of glacial acetic acid and 3 cm³ of phosphorus trichloride into the flask; immediately attach the condenser and shake and cool the flask.

3 Set up the apparatus as in figure 4.D attaching a small absorption tube containing loosely packed anhydrous calcium chloride to the side arm of the receiver to prevent moisture entering the apparatus. Heat the flask on a water bath, carefully at first because the evolution of hydrogen chloride may be vigorous, then bring the water up to its boiling-point.

4 Allow the apparatus to cool for five minutes and then drain away the residue in the reaction flask (do not wash out the flask). Put the crude acetyl chloride from the receiver in the flask and redistil the acetyl chloride collecting the portion boiling at 324 ± 2K (51 ± 2°C).

4.40 The Properties of Acid Chlorides and Anhydrides.

Acetyl (ethanoyl) chloride and acetic (ethanoic) anhydride may be compared.

1 Add 0.5 cm³ of acetyl chloride to 3 cm³ of water in a test-tube in the fume cupboard. Add 1 cm³ dilute nitric acid and then a few drops of silver nitrate solution.

2 Add 0.5 cm³ of acetic anhydride to 3 cm³ of water in a test-tube in the fume cupboard: note the appearance in the cold and then on warming.

3 A method of esterification of both liquids is as follows: carefully add 1 cm³ of the liquid to 1 cm³ of ethanol and then gently warm the mixture. Pour the contents of the tube into 5 cm³ of water in a small beaker and then add sodium carbonate solution until the solution is alkaline. Note the odour of the ester.

4 To 2 cm³ of aniline (phenylamine) in a flask very carefully add 1 cm³ of acetyl chloride. Then add sufficient water (about 15 cm³) to dissolve the solid on boiling; crystals of acetanilide (N-phenylacetamide) separate on cooling.

5 When test (4) is repeated 2 cm³ of acetic anhydride should be used and the mixture heated on a water-bath for five minutes: this gives a good illustration of the relative reactivity of the substances.

6 To 0.5 cm³ of acetic anhydride add one spatula load of hydroxyammonium chloride and 3 cm³ of sodium hydroxide solution, boil the mixture and then cool it before acidifying with dilute hydrochloric acid. Hydroxamic acid is present in the solution and when a few drops of iron(III) chloride solution are added a wine-red coloration is obtained, e.g.

$$(CH_3CO)_2O + NH_2OH \rightarrow CH_3CONHOH + CH_3COOH$$
Hydroxamic acid
(N-hydroxyethanamide)

4.41 The Preparation of Benzoyl Chloride

Care: this experiment must be done in the fume-cupboard. Benzoyl chloride is lachrymatory (makes your eyes water). The experiment illustrates the replacement of —OH in an acid by a —Cl group.

1 Set up the apparatus as in figure 4.D: it must be very carefully dried. Into the flask put 20 cm³ of thionyl chloride (sulphur dichloride oxide) together with a few anti-bumping granules and carry out a distillation collecting the portion boiling at 352 ± 2K (79 ± 2°C).

2 Tip the residue in the distillation flask down the fume cupboard drain with plenty of water and allow the flask to drain for a moment. Rearrange the apparatus to be as in figure 4.B. Into the flask put 10 g of dry powdered benzoic acid and add 10 cm³ of the redistilled thionyl chloride. Heat the flask on a boiling water bath, with occasional shaking, for 20 minutes or until the evolution of hydrogen chloride ceases.

3 Allow the flask to cool for five minutes before re-arranging the apparatus to be as in figure 4.D again; an absorption tube containing loosely packed anhydrous calcium chloride should be fitted to the receiver flask to prevent the entry of moisture. Heat the flask over a gauze, discard the first fraction which is mainly unchanged thionyl chloride, then turn off the water condenser and change the receiver to collect the fraction boiling at 470 ± 2K (197 ± 2°C).

4.42 The Preparation of Acetamide (Ethanamide)

1 Set up the apparatus for reflux distillation (figure 4.A). Into the flask put 15 g of ammonium acetate (ethanoate) and 15 cm³ of glacial acetic (pure ethanoic) acid together with a few anti-bumping granules. Boil **the mixture for at least two, preferably four, hours.**

2. Rearrange the apparatus for distillation (figure 4.C). Heat the mixture in the flask and distil off acetic acid and water. When the temperature reaches 423 K (150 °C) stop heating the flask, allow it to cool for a few minutes and then take out the water condenser. Reheat the mixture and collect the acetamide that distils in the range 494 ± 2 K (221 ± 2 °C). Some may solidify in the air condenser: melting-point 355 K (82 °C).
$$CH_3COONH_4 \rightarrow CH_3CONH_2 + H_2O$$
The usual odour of acetamide is that of mice; it is due to an impurity which can be removed by redistillation or recrystallization.

4.43 The Preparation of Benzamide

1 Into the flask put 10 cm³ of concentrated ammonia solution and 10 cm³ of water. In a fume cupboard add 2 cm³ of benzoyl chloride (**care**: it is lachrymatory). With the thermometer pocket (for the Quickfit apparatus 29 BU) or a well-fitting stopper in the top of the flask shake the flask for about 15 minutes. From time to time release the stopper so that any excess pressure generated can be counteracted. After a while no trace of the oily benzoyl chloride should be seen.

2 Filter off the crystals of benzamide on a small Hirsch funnel and wash them with cold water. Recrystallize the benzamide from hot water.

4.44 The Properties of Ammonium Salts and Amides

Ammonium acetate (ethanoate) and acetamide (ethanamide) may be compared.

1 To 2 cm³ of water slowly add some of the solid until some remains undissolved. Keep the solution for test (4).

2 To half a spatula load of the solid add 2 cm³ of sodium hydroxide solution. If there is no gas evolved **in the cold, warm the mixture gently.**

3 To half a spatula load of the material in the flask add about 10 cm³ of sodium hydroxide solution and reflux the mixture gently over a Bunsen burner and gauze for 15 minutes. Cool the flask and acidify the mixture with dilute sulphuric acid. A white precipitate would indicate an aromatic acid but in its **absence the solution should be tested for acetic acid, etc. It may be useful to distil off the acetic acid in order to identify it.**

4 Cool 1 cm³ of a concentrated solution of sodium nitrite [nitrate(III)] and then add 1 cm³ of dilute hydrochloric acid. Allow any reaction to subside before adding the solution, which effectively consists of nitrous [nitric(III)] acid, to the cooled solution from test (1).

5 To a spatula load of acetamide (only) add, **with great care,** four drops of bromine and 3 cm³ of sodium hydroxide solution. Cork the test-tube and shake it for three minutes. Remove the cork and add two pellets of sodium hydroxide. Gently boil the mixture and note the odour and effect on damp litmus paper of the gas evolved.
Safety spectacles must be worn; even so, this experiment should be done in a fume cupboard.

4.45 The Preparation of Ethyl Acetate (Ethanoate)

This preparation illustrates the process of esterification.

1 Carefully put 6 cm^3 of ethanol and 4 cm^3 of glacial acetic (pure ethanoic) acid into the flask and fit it with a reflux condenser. Allow the liquids to mix thoroughly before removing the condenser, cautiously pouring in 2 cm^3 of concentrated sulphuric acid and **replacing the condenser. Next the flask should be** cooled in a beaker of water until the liquids have mixed thoroughly. Then heat the beaker until the water boils and allow the reagents to reflux for 30 minutes (figure 4.A).

2 Allow the flask to cool for five minutes before re-arranging the apparatus for distillation (figure 4.C). Heat the flask on a gauze over a small flame and collect the distillate in the receiver: continue until about 5 cm^3 of distillate have formed or the contents of the flask show signs of the evolution of a gas (ethene).

3 Wash the distillate with 2 cm^3 of sodium carbonate solution, then 2 cm^3 of water and finally three times with 2 cm^3 portions of a concentrated solution of calcium chloride. State why each of these processes is necessary. A teat pipette is useful to ensure thorough mixing of these wash-solutions with the crude ethyl acetate: they may be ejected from the pipette into the organic layer. Dry the ethyl acetate by allowing it to stand for 30 minutes in contact with a few pieces of anhydrous calcium chloride.

4 Redistil the ethyl acetate collecting the portion boiling at $330 \pm 2\,K$ ($57 \pm 2\,°C$).

4.46 The Preparation of Ethyl Benzoate

This illustrates the process of esterification by the Fischer-Speier method.

1 In a fume cupboard set up a modified version of the apparatus in figure 4.B. To the outer end of the jet tube attach a hydrogen chloride generator: drip concentrated sulphuric acid into concentrated hydrochloric acid in a conical or a round-bottomed flask. To the inner end of the jet tube attach a fine piece of plastic tubing extending down into the flask. To the end of the adaptor tube, hanging downwards, attach an absorption tube loosely filled with anhydrous calcium chloride. The flask should be heated on a sand-bath.

2 Disconnect the flask and put 5 g of benzoic acid and 5 cm^3 of ethanol in the flask. Pass hydrogen chloride directly into the mixture for a few minutes and then put the flask back under the reflux condenser. The supply of hydrogen chloride must be rapid during the initial stages of the experiment because it is very soluble in ethanol and sucking back must not be allowed to occur.

   Heat the flask and maintain the supply of hydrogen chloride; the supply can be diminished when the solution reaches its boiling-point. Boil the mixture for an hour.

3 Allow the apparatus to cool for a few minutes and then cool the flask with tap-water. Pour the solution into a separating funnel containing about 50 cm^3 of water. Rinse out the flask with 5 cm^3 of water and pour the rinsings into the funnel. To facilitate the separation of ethyl benzoate from water (the densities are 1.05 and 1.00 g/cm^3 respectively) add 3 cm^3 of carbon tetrachloride (tetrachloromethane). Run off the lower layer which is of ethyl benzoate in carbon tetrachloride and discard the upper layer.

   Return the ethyl benzoate to the funnel and shake it with an equal volume of sodium carbonate solution to neutralize any acid; add 5 cm^3 more of sodium carbonate solution and if effervescence again occurs continue in this manner until there is no effervescence. Run off the lower layer into a conical flask and dry it with anhydrous calcium chloride for 15 minutes.

4 Decant or filter the dried ester back into the flask. Add a few anti-bumping granules and set up the apparatus for distillation as in figure 4.D. Heat the flask to distil off the carbon tetrachloride (boiling-point 350 K or 77 °C) and then drain the condenser. Continue heating the flask and collect the ethyl benzoate fraction at $486 \pm 2\,K$ ($213 \pm 2\,°C$).

4.47 The Preparation of Phenyl Benzoate

This experiment illustrates the process of benzoylation: the Schotten-Baumann reaction. The first stage must be confined to the fume cupboard because benzoyl chloride is lachrymatory. **Care: phenol is corrosive.**

1 Carefully put two spatula loads of phenol into the flask and add 15 cm^3 of sodium hydroxide solution; shake until the phenol has dissolved. Next add 2 cm^3 of benzoyl chloride, cork the flask and shake it for 15 minutes. The ester separates as a solid and can be filtered off: any lumps should be carefully broken up using a glass rod and the ester washed thoroughly to eliminate any remaining benzoyl chloride.

2 The crude ester should be recrystallized from ethanol. Put the ester in the flask and just cover it with ethanol. Set up the apparatus as in figure 4.A and heat the flask until the ethanol boils. More ethanol may be added down the condenser if it is necessary for dissolving the ester but care must be taken not to add an excess at this stage.

   In order to permit the separation of the solid ester rather than the liquid (melting-point 342 K, 69°C), the volume of ethanol must be nearly doubled. When the appropriate quantity of ethanol has been added heat the solution to boiling again and then filter it.

4.48 The Properties of Esters

1 Put 2 cm^3 of ethyl acetate (ethanoate) in the flask and add 10 cm^3 of concentrated sodium hydroxide solution, 10 cm^3 of water and a few anti-bumping granules. Set up the apparatus as in figure 4.A. Heat the mixture over a tripod and gauze for 20 minutes. Using an alkali is usually the most rapid way of hydrolyzing an ester.

   Allow the apparatus to cool for five minutes before rearranging it for distillation as in figure 4.C. Reheat the mixture and collect about 4 cm^3 of distillate: this is water and ethanol. Demonstrate the presence of ethanol by performing the iodoform reaction [see experiment 4.21 (4)].

   Allow the apparatus to cool again and then add, with stirring, dilute sulphuric acid until the mixture is definitely acidic: this liberates acetic (ethanoic) acid into the solution. Distil off about 4 cm^3 of the mixture: water distils as well as acetic acid. Demonstrate the presence of acetic acid in the distillate (see experiment 4.34).

2 Esters will also yield hydroxamic acid: repeat experiment 4.40 (6) using ethyl acetate instead of acetic anhydride.

3 Many methyl and ethyl esters react with benzylamine [(phenylmethyl)amine], in the presence of ammonium chloride as a catalyst, to give crystalline N-benzylamides. Into the flask put 1 cm^3 of ethyl acetate, 3 cm^3 of benzylamine, a spatula load of ammonium chloride and a few anti-bumping granules. Set up the apparatus as in figure 4.A and allow the mixture to reflux for 15 minutes. Allow the apparatus to cool for five minutes before adding 10 cm^3 of water to dissolve any excess of benzylamine. Filter off the benzylamide.

4.49 The Hydrolysis of Phenyl Benzoate

1 Put 2 cm^3 of phenyl benzoate in the flask and add 10 cm^3 of concentrated sodium hydroxide solution, 10 cm^3 of water and a few anti-bumping granules. Set up the apparatus as in figure 4.A. Heat the mixture over a tripod and gauze for 20 minutes. The solution remaining has to be treated differently than that for simple esters because neither sodium phenolate (phenoxide) nor sodium benzoate are volatile.

2 Cool the flask and its contents thoroughly before adding dilute sulphuric acid, with stirring, until the solution is definitely acidic. Benzoic acid starts to separate and is redissolved by adding sodium carbonate solution, with vigorous stirring, until the solution is definitely alkaline. The solution is then treated twice with 15 cm^3 portions of diethyl ether (ethoxyethane)· both the water and the ether layers of the solution must be kept. **Care: extinguish all flames nearby. Phenol dissolves in the ether.**

3 The solution of the phenol in ether is put in the flask and the apparatus set up for distillation as in figure 4.D. Heating of the flask is carried out best by means of hot water. A rubber tube may be fitted to the flask so that any fumes are carried well away. The residue remaining in the flask is of phenol and can be investigated as in experiment 4.23.

4 The solution of sodium benzoate in water should be acidified again with dilute sulphuric acid: benzoic acid separates as a white precipitate. Filter off the acid, wash it well with water; its properties are investigated in experiment 4.38.

Nitrogen and Sulphur Compounds

4.50 The Preparation of Nitrobenzene

This preparation is done to illustrate the process of nitration; it involves the substitution of a hydrogen attached to the benzene ring by an $-NO_2$ group.

1 Put 8 cm³ of concentrated nitric acid into the flask together with a few anti-bumping granules. Next carefully add 8 cm³ of concentrated sulphuric acid. In a fume cupboard add four 2 cm³ portions of benzene to the mixed acid: have a thermometer in the flask and be prepared to cool the flask in a beaker of cold water if the temperature rises above 323 K (50°C). When the three reagents are all fully mixed fit the flask with a reflux condenser (figure 4.A) and raise the temperature of the water in the beaker to 333 K (60°C): keep it so for 30 minutes, swirling the flask from time to time to ensure thorough mixing.

What is the function of the concentrated sulphuric acid? Write molecular and ionic equations for the reaction.

2 Remove the beaker of water and allow the apparatus to cool for five minutes before pouring the contents of the flask into 75 cm³ of cold distilled water in a beaker: stir well during the addition. Decant off as much as possible of the water layer and repeat the washing with about 75 cm³ of distilled water. Then wash the oily layer with 75 cm³ of sodium carbonate solution and finally with 75 cm³ of distilled water again. Dry the nitrobenzene by leaving it in contact with some granules of anhydrous calcium chloride for 20 minutes.

3 Set up the apparatus (figure 4.D) for distillation. Decant the nitrobenzene into the flask. Water should be drained from the condenser after any remaining benzene has distilled off (about 353 K, 80 °C). Collect the fraction distilling at 484 ± 2 K (211 ± 2 °C).

4.51 The Preparation of 1,3-Dinitrobenzene

This is done to illustrate the more difficult process of further nitration; the nitro group has a deactivating effect upon the benzene nucleus.

1 Put 4 cm³ of concentrated nitric acid into the flask together with a few anti-bumping granules. Carefully add 5 cm³ of concentrated sulphuric acid and then set up the apparatus as in figure 4.B. Add three 1 cm³ portions of nitrobenzene at minute intervals, swirling the flask continuously. Heat the flask in boiling water for 40 minutes, shaking the flask at five minute intervals to ensure good mixing.

2 Remove the water bath and allow the apparatus to cool for five minutes before pouring the contents of the flask into 75 cm³ of cold distilled water, stirring continuously to prevent any lumps forming as the product solidifies. Filter off the solid on a Hirsch funnel, pressing the crystals with a glass stopper or a clean cork. Wash the crystals with cold distilled water, allow them to drain and then press them between filter-papers to dry them.

3 The crude product can be recrystallized from ethanol. The solid is put in a beaker and 25 cm³ of ethanol added. The mixture is heated until the solid dissolves. The beaker is then cooled in water, stirring the solution to prevent the formation of lumps. When crystallization is complete the product can be filtered off on a Hirsch funnel and dried between filter-papers. If crystallization is too slow add water dropwise to the ethanolic solution. Colourless crystals should be obtained: their melting-point may be determined to **assess their purity (363 K, 90 °C).**

4.52 The Properties of Aliphatic Amines

Methylamine, ethylamine, propylamine, butylamine, diethylamine or triethylamine may be studied.

1 Shake 2 cm³ of water with a few drops of the amine. Test the water (a solution?) with pH paper.

2 Put a few drops of the amine on a crucible lid and ignite it with a lighted splint.

3 Ascertain whether a few drops of the amine mix with a 2 cm³ portion of dilute hydrochloric acid.

4 To 2 cm³ of water add 2 cm³ of concentrated hydrochloric acid and cool the moderately concentrated hydrochloric acid so formed in ice. Add five drops of the amine and then eight drops of cold concentrated sodium nitrite [nitrate(III)] solution.

The evolution of nitrogen indicates the presence of a primary amine. The separation of an oil or formation of a turbid solution indicates a secondary amine. If the solution does not appear to react it is of a tertiary amine although dimethylaniline gives 4-nitrosodimethylaniline.

5 Primary amines give the isocyanide test but this is not recommended [see experiment 4.54 (6)] because the aliphatic isocyanides are very poisonous.

4.53 The Preparation of Aniline (Phenylamine)

This is done to illustrate the reduction of a nitro group, ether extraction and steam distillation.

1 Put 3 cm^3 of nitrobenzene and about 7 g of granulated tin into the flask and set up the apparatus as in figure 4.B. Being prepared to cool the flask in a beaker of water if the reaction becomes too vigorous, add four 3 cm^3 portions of concentrated hydrochloric acid, swirling the flask to ensure thorough mixing. Then heat the water in the beaker and maintain it at its boiling-point for 20 minutes.

2 Remove the water bath and cool the flask with cold water before adding a very concentrated solution of sodium hydroxide in water (9 g solid in 12 cm^3 water). Set up the apparatus for steam distillation (figure 4.E) generating the steam in a conical flask fitted with a safety tube. Heat the conical flask and the flask containing the reaction mixture until distillation starts; after that heating the conical flask only should suffice. Continue distillation until about $20–25 \text{ cm}^3$ of distillate have collected in a beaker.

3 Stir about 4 g of sodium chloride, in the form of fine crystals, into the distillate. In the absence of any flames add 6 cm^3 of diethyl ether (ethoxyethane) and mix it thoroughly with the water layer using a teat pipette to inject the ether into the water. Allow the mixture to separate into two layers and transfer the upper ethereal one to a boiling tube ($150 \times 25 \text{ mm}$). Repeat the extraction on the water layer with a second portion of ether and add this extract to the first. Dry the ether extracts by adding four pellets of sodium hydroxide and allowing the mixture to stand for 20 minutes. **Care is vital in this stage.**

4 Decant the ether extract into a dried flask containing a few anti-bumping granules and set up the apparatus for distillation (figure 4.D). Away from the distillation apparatus, heat water in a beaker to its boiling-point. Carefully raise the beaker to heat the flask and distil off the ether at $308 \pm 2 \text{K}$ ($35 \pm 2 \,^{\circ} \text{C}$). The flask must not be heated with a naked flame. When the ether has distilled off change the receiver, cease to pass water through the condenser and collect the fraction boiling at $457 \pm 2 \text{K}$ ($184 \pm 2 \,^{\circ} \text{C}$).
   Write equations for the reactions involved.

4.54 The Properties of Aniline (Phenylamine)

1 Shake 2 cm^3 of water with a few drops of aniline. Test the water (a solution?) with pH paper.

2 Put a few drops of aniline on a crucible lid and ignite it with a lighted splint.

3 To two drops of aniline add four drops of water and note the appearance of the mixture before and after shaking. Next add six drops of concentrated hydrochloric acid and again note the appearance. Finally add sodium hydroxide solution until the mixture is alkaline and once again note the appearance.

4 Into a dry test-tube put 0.5 cm^3 of aniline and 0.5 cm^3 of acetic anhydride. Cool the tube, then pour the reaction mixture into about 10 cm^3 of distilled water in a beaker. What do you observe? See also section 4.55.

5 The benzoylation of aniline is carried out in a similar manner to acetylation but the use of benzoyl chloride, as in the Schotten-Baumann method, means that the reaction must be done in a fume cupboard because the reagent has an irritant vapour. Into a boiling tube place 0.5 cm^3 of aniline and 0.75 cm^3 of benzoyl chloride followed by about 15 cm^3 of sodium hydroxide solution. Fit the tube with the stopper and shake vigorously for five minutes. Note the appearance of the residue. See also section 4.56.

6 The isocyanide (carbylamine) test for primary amines must be carried out in a fume cupboard and the products confined there. To **one drop** of aniline add **one drop** of chloroform (trichloromethane) and then add some potassium hydroxide dissolved in 1 cm^3 of ethanol. Warm the mixture carefully and sample the vapours very cautiously (poison). Cool the mixture and then add 5 cm^3 concentrated hydrochloric acid. Pour the hydrolyzed isocyanide down the fume cupboard drain and rinse out the test-tube with another portion of acid.

7 Shake one drop of aniline with 5 cm^3 of water in a test-tube and then add a few drops of sodium hypochlorite [chlorate(I)] solution (or a suspension of bleaching powder in water). Note the appearance.

8 Shake two drops of aniline with 4 cm^3 of water in a test-tube in a fume cupboard, then slowly add bromine water. Note the appearance of the reaction mixture.

9 To 2 cm^3 of water in a test-tube add 2 cm^3 of concentrated hydrochloric acid and then 0.5 cm^3 of aniline; divide the solution into two portions. Into another test-tube put 1 cm^3 of water and stir in three spatula loads of sodium nitrite [nitrate(III)]; divide the solution into two portions. Allow one pair of solutions to react at room temperature, noting the appearance of the reaction mixture.
   Cool the second pair of solutions in ice and mix them slowly with stirring, keeping the temperature below 280K ($7 \,^{\circ} \text{C}$). Again note the appearance. See also experiment 4.58.

4.55 The Preparation of Acetanilide (N–phenylethanamide)

This illustrates the process of acetylation. The amine group in aniline (phenylamine) is very susceptible to oxidation, hence all samples of aniline except those freshly distilled are coloured. The formation and hydrolysis of acetanilide may be used to purify aniline. Acetanilide formed by 'blocking' the amine group may be hydrolyzed after reactions have been carried out elsewhere in the benzene ring, e.g. in the preparation of 4-nitroaniline.

1 Into a dried flask put 1 cm^3 of aniline together with a few anti-bumping granules and cautiously add 1.5 cm^3 of acetic anhydride with gentle swirling to ensure thorough mixing. Attach the flask to a reflux condenser, as in figure 4.A, and heat the flask in a beaker of boiling water for 15 minutes.

2 Remove the beaker and allow the apparatus to cool for five minutes before pouring the contents into about 60 cm^3 of cold distilled water in a beaker. Then heat the beaker until the crude acetanilide dissolves. Cool the beaker in water and filter off the crystals using a Hirsch funnel, Buchner flask and filter-pump. Wash the crystals with cold distilled water and allow them to drain. The crystals may be pressed with a glass stopper or a clean cork to assist the removal of water. Finally the crystals may be dried by pressing them between filter-papers.

4.56 The Preparation of Benzanilide (N–phenylbenzamide)

Benzoylation of aniline (phenylamine) may be carried out for similar reasons to acetylation – see experiment 4.55. In the Schotten-Baumann reaction benzoyl chloride is the reagent, the vapour of which is irritant and may be lachrymatory, and so its use must be restricted to the fume cupboard.

1 Into the flask put 1 cm^3 of aniline and about 25 cm^3 of sodium hydroxide solution. Carefully add 1.5 cm^3 of benzoyl chloride, gently swirling the flask to ensure thorough mixing. Put the stopper on the flask and shake vigorously for 15 minutes. Note the appearance of the reaction mixture. Filter off the ester on a Hirsch funnel, breaking up any lumps; wash it thoroughly with distilled water and allow to drain.

2 The crude amide can be recrystallized from ethanol: put the amide back into the flask, just cover it with ethanol, add a few anti-bumping granules and set up the apparatus for refluxing, as in figure 4.A. Heat the flask until the ethanol boils, add more ethanol down the condenser until the ester just dissolves and then add an excess of 5 cm^3. In a beaker heat some ethanol which is then poured through a Hirsch funnel to warm it. Filter the hot solution of benzanilide through the hot funnel. Allow the filtrate to crystallize and collect the crystals by a second filtration.

4.57 The Preparation of Benzylidene-Aniline (N–Benzylidenephenylamine)

This compound is an example of a Schiff's base and this reaction may be used to characterize aldehydes, particularly aromatic ones.

1 Into a crucible resting on a small beaker of boiling water put 1 cm^3 of benzaldehyde and 1 cm^3 of aniline (phenylamine). Stir the mixture with a glass rod at intervals during the next 20–30 minutes. Then remove the beaker and allow the crucible to cool for five minutes before standing it on a piece of ice; stir the mixture well.

2 Chop up the crude crystalline product in the crucible and put it into the flask. Recrystallize it from ethanol as in experiment 4.56(2).

4.58 The Properties of Benzenediazonium Chloride

1 The benzenediazonium chloride should be prepared by scaling up the quantities in experiment 4.54 (9) by a factor of five: put 10 cm³ of water in a beaker, add 10 cm³ of concentrated hydrochloric acid and then 2.5 cm³ of aniline. Into a boiling tube put 5 cm³ of water and stir in about 3 g of sodium nitrite [nitrate(III)] . Cool the beaker and the boiling tube in ice and salt. Slowly pour the sodium nitrite solution into the anilinium (phenylammonium) chloride solution, stirring well and not allowing the temperature to rise above 280K (7°C). The solution is used in the four tests following.

2 Warm a sample of the benzenediazonium chloride solution. Nitrogen is evolved. The odour of a second product is characteristic.

3 To a second sample add 1 cm³ of dilute sulphuric acid and then 1 cm³ of potassium iodide solution. Warm the mixture; nitrogen is again evolved. Note the appearance on subsequent cooling.

4 Dissolve a spatula load of phenol in sodium hydroxide solution in a beaker. Add some of the benzenediazonium chloride solution and note the appearance of the product. **Care:** the phenolic solution is corrosive.

5 Dissolve a spatula load of resorcinol (benzene-1, 3-diol), in sodium hydroxide solution in a beaker. Add some of the benzenediazonium chloride solution and note the appearance of the product. The dye produced is known as Resorcin red.

4.59 The Preparation of Methyl Orange

This is an azo-dye and the preparation is similar to that in experiments 4.58 (4) and (5).

1 Put six spatula loads of sulphanilic acid (4-amino-benzenesulphonic acid) and two of anhydrous sodium carbonate in about 25 cm³ of warm distilled water in a beaker. Add more sodium carbonate if necessary so that the mixture is alkaline to litmus. Add two spatula loads of sodium nitrite [nitrate(III)] and cool the mixture below 280K (7°C).

2 Put 8 cm³ of water in a boiling tube and add 4 cm³ of concentrated hydrochloric acid. Cool the moderately concentrated hydrochloric acid so produced and add it slowly with stirring to the mixture in the beaker, not allowing the temperature to rise above 283K (10°C).

3 Allow the mixture from (2) to stand whilst dissolving 2 cm³ of dimethylaniline (N,N-dimethylphenylamine) in 2 cm³ of concentrated hydrochloric acid and 5 cm³ of water. Cool the dimethylanilinium chloride solution to below 280K (7°C) and then add it slowly with stirring to the mixture from (2).

4 After allowing the reaction in step (3) to proceed for a few minutes add sodium hydroxide solution and note the appearance of the product.

5 The product is slightly soluble in water and the solution may be tested with alkalis and with acids in dilute solution.

4.60 The Properties of Urea (Carbamide)

1 Shake 2 cm³ of water with a few urea crystals. Test the water (a solution?) with pH paper.

2 Put a few urea crystals on a crucible lid and ignite them with a lighted splint.

3 Put two spatula loads of urea in a dry test-tube and carefully heat the urea to just above its melting-point; identify the gas evolved. Note the subsequent behaviour of the liquid and then stop heating.

   Dissolve the solid residue in 4 cm³ of sodium hydroxide solution, cool the solution and then add one drop of a very dilute solution (about 0.01M) of copper(II) sulphate. Note the coloration: it is often given by compounds containing –CO–NH– groups.

4 Put one spatula load of urea into 3 cm³ of sodium hydroxide solution in a test-tube and boil the mixture. Identify the gas evolved.

5 Put one spatula load of urea into 3 cm³ of dilute sulphuric acid in a test-tube and boil the mixture. Identify the gas evolved.

6 To 2 cm³ of bromine water add sufficient sodium hydroxide solution to cause decolorization. Then sprinkle in a few crystals of urea and identify the gas(es) evolved.

7 Add urea to 2 cm³ of water until no more will dissolve (at room temperature) and then divide the solution into two portions: to one add a few drops of concentrated nitric acid and to the other add a few drops of a concentrated solution of oxalic (ethanedioic) acid.

4.61 The Preparation of Nylon 66

This demonstration is very expensive but it illustrates the process of condensation polymerization.

1 Make up 5% m/V solutions of hexamethylenediamine (hexane-1,6-diamine) in distilled water and of adipyl chloride (hexanedioyl dichloride) in carbon tetrachloride (tetrachloromethane). Put the adipyl chloride solution in a small beaker and then carefully add the amine solution so that the latter forms an upper layer.

2 A skin is formed between the two layers; with a glass rod to the centre of the skin pull out the solid that forms at the interface of the solutions and wind the fibre on to the rod. The fibre can be stretched and will then show some elasticity. The solutions can also be stirred to yield a lump of nylon.

4.62 The Preparation of Sodium Alkylbenzene-sulphonate

This is a common soapless detergent: the alkyl group is twelve carbon atoms long and sulphonation of the benzene ring occurs at position four. Dodecylbenzene can be obtained from Unilever Education Section, Unilever House, Blackfriars, London EC4P 4BQ.

The experiment should be carried out in a fume cupboard: rubber gloves, safety-glasses and an overall being worn by the student. The quantities should be measured out accurately in a measuring cylinder.

The sulphonation should be carried out in a wide-necked round-bottomed flask or a small beaker (about $100 \, cm^3$) held in a water (or ice) bath.

---

1   Put $23 \, cm^3$ of dodecylbenzene into the reaction vessel and ice into the water bath. Cool the vessel and contents to below 283 K ($10°C$). From the measuring cylinder slowly add $10 \, cm^3$ of oleum (sulphuric acid containing 20% free sulphur trioxide): vigorous stirring is essential and the rate of addition must be such that the temperature never rises above 328 K ($55°C$). It may be necessary to add more ice to the bath. The temperature should be allowed to rise to 328 K at the end of the last addition. The ice is then replaced by hot water and the temperature should be maintained at 328 K ($55°C$) for 30 minutes. Then the hot water should be replaced by some more iced water and the temperature of the vessel and contents reduced to below 293 K ($20°C$).

---

2   To neutralize the benzenesulphonic acid prepared in (1) a solution of 3.2 g of sodium hydroxide in $6 \, cm^3$ of water is prepared. Place a $100 \, cm^3$ beaker surrounded by ice in the bath. Into the beaker put 8 g of finely crushed ice and then the concentrated sodium hydroxide solution. Add about a quarter of the sulphonic acid to the alkali with vigorous stirring: the temperature must not be allowed to rise above 323 K ($50°C$) during the neutralization. Then by addition of more benzenesulphonic acid or dilute sodium hydroxide solution adjust the pH to between 6 and 8; the closer to 7 the more satisfactory the product will be.

---

3   Shake small samples of the soapless detergent with
    a) distilled water
    b) tap-water
    c) temporarily hard water
    d) permanently hard water.

4.63 The Properties of Amino Acids

A typical amino acid is glycine (aminoethanoic acid), H_2NCH_2COOH.

1 Compare the solubility of glycine in 1 cm^3 portions of water, dilute hydrochloric acid and dilute sodium hydroxide solution. What is the pH of its aqueous solution? Keep this solution for (2).

2 To the aqueous solution add a spatula load of sodium carbonate: at what rate is carbon dioxide evolved in the cold and on warming?

3 **Van Slyke's estimation of nitrogen present as an amine**
To 2 cm^3 of sodium nitrite [nitrate(III)] solution add 0.5 cm^3 of dilute acetic (ethanoic) acid. Add this solution, which contains nitrous acid, to a cold aqueous solution of glycine. Identify the gas evolved.

4 Mix one spatula load of glycine with one of soda-lime. Heat the mixture and note the evolution of methylamine.

5 To 2 cm^3 of glycine solution add a few drops of a very dilute solution (about 0.01M) of copper(II) sulphate followed by an excess of sodium hydroxide solution. Note the coloration; compare 4.60 (3) and 4.64 (1).

6 **Electrophoresis**
Take a piece of slotted filter-paper or four 1 cm wide strips of filter-paper and suspend them in a Λ-formation over a glass rod with 0.5 cm between each. The strip of filter-paper to be used must be clean and should be handled only with tweezers. Soak the four strips with buffer solutions having pH's of 3, 6, 7 and 9 respectively (see experiment 1.54). Arrange the ends of the strips to dip into two beakers containing sodium sulphate solution. Put saturated glycine solution centrally on each of the strips. Carbon rods should be placed in the beakers and a potential of 60 volts applied. The assembly should be kept free from draughts by a large inverted beaker and the potential applied for about four hours.

Then dry the paper before dipping it into or spraying on ninhydrin [0.25% solution in acetone (propanone)]. The reaction between ninhydrin and the amino-acid is accelerated by placing the paper in an oven set at about 350K (about 77°C) for approximately half an hour. The bluish-purple spots obtained should be ringed with a pencil because they may fade later.

The glycine will be found to have moved towards the cathode or anode or not at all. The pH at which it has not moved at all is the iso-electric point. What is this pH for glycine? Explain the three possibilities.

4.64 Some Tests on Proteins

Bread, milk, egg (white and yolk separately), flour, fish, meat, peanuts, etc. may be tested.

1 The Biuret test

To $2\,cm^3$ of a solution or a suspension in water of the substance add one drop of a very dilute solution (about 0.01M) of copper(II) sulphate followed by an excess of sodium hydroxide solution. Note the coloration: it is given not only by proteins but also by many of their products upon hydrolysis. A pink or violet coloration is often given by compounds containing two or more peptide (−CO−NH−) bonds.

2 Xanthoproteic test

To $2\,cm^3$ of a solution of the substance add two drops of concentrated nitric acid, gently warm the mixture and then add an excess of ammonia solution. The orange colour indicates a protein containing amino acids such as tyrosine, tryptophan and phenylalanine.

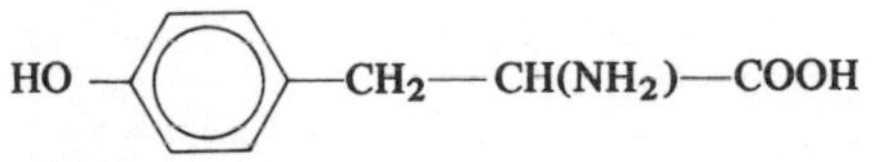

$$HO-\bigcirc-CH_2-CH(NH_2)-COOH$$
Tyrosine

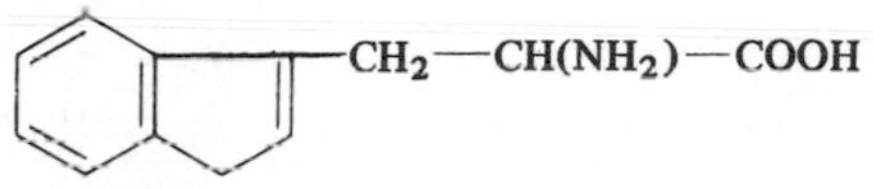

$$CH_2-CH(NH_2)-COOH$$
Tryptophan

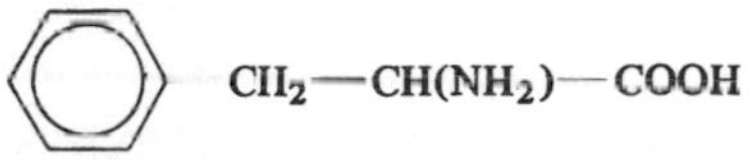

$$CH_2-CH(NH_2)-COOH$$
Phenylalanine

3 Millon's test

To $2\,cm^3$ of a solution of the substance add mercury(I) nitrate solution containing nitrous [nitric(III)] acid (made by a few crystals of sodium nitrite [nitrate(III)] and two drops of dilute nitric acid). Note the immediate result and the coloration obtained on standing. A white precipitate which later turns pink is given by a protein.

4 Lead(II) acetate test

To $2\,cm^3$ of a solution of the substance add $0.5\,cm^3$ of sodium hydroxide solution and then sufficient acetic (ethanoic) acid to make the solution just acidic to litmus. Finally add a few drops of lead(II) acetate (ethanoate) and note the result. A black precipitate is given by proteins containing methionine and cystine.

$$CH_3-S-CH_2-CH_2-CH(NH_2)-COOH$$
Methionine

$$S-CH_2-CH(NH_2)-COOH$$
$$|$$
$$S-CH_2-CH(NH_2)-COOH$$
Cystine

5 Glyoxylic acid test

(HOOC−CHO). To $2\,cm^3$ of a solution or a suspension in water of the substance add a few crystals of glyoxylic (oxoethanoic) acid. Holding the test-tube at an angle of $45°$ carefully pour in a little concentrated sulphuric acid. Note the coloration at the boundary between the two solutions. A violet coloration is given by proteins containing tryptophan.

6 Sakaguchi's test

To $2\,cm^3$ of a solution of the substance add a crystal of naphth-1-ol and then a few drops of sodium hypochlorite [chlorate(I)]. Note the coloration. A violet coloration is given by proteins containing arginine.

$$HN=C-(CH_2)_3-CH-COOH$$
$$\ \ \ |\qquad\qquad\quad |$$
$$\ \ NH_2\qquad\quad NH_2$$
Arginine

4.65 The Properties of Carbohydrates

Glucose, fructose, sucrose and starch may be studied.

1 Add a spatula load of the substance to 3 cm^3 of water. Shake well, and warm if no effect in the cold.

2 Heat a spatula load of the substance in an ignition tube. Are carbon dioxide and/or steam evolved?

3 Molisch's test: dissolve half a spatula load of the substance in 3 cm^3 of water, add three drops of naphth-1-ol in water and ethanol (4:1) then carefully pour in 2 cm^3 of concentrated sulphuric acid so that it forms a lower layer.

4 Warm half a spatula load of the substance with 1 cm^3 of concentrated sulphuric acid using a small flame.

5 Boil half a spatula load of the substance with 4 cm^3 of sodium hydroxide solution.

6 Fehling's solution – see section 4.27(3).

7 Tollen's reagent – see section 4.27(4).

8 Rapid furfural test: boil together 0.5 cm^3 of a 1% solution of naphth-1-ol in ethanol and 4 cm^3 of concentrated hydrochloric acid.

9 Osazone formation: dissolve half a spatula load of the substance in 2 cm^3 of water, add five drops of 2,4-dinitrophenylhydrazine solution [in 3M hydrochloric acid, Brady's reagent] and five drops of glacial acetic (ethanoic) acid, and warm in a beaker for at least 20 minutes.

10 Hydrolysis: in a small beaker dissolve a spatula load of sucrose in 6 cm^3 of water and add 4 cm^3 dilute sulphuric acid. Boil the solution for five minutes and then allow the solution to cool. Neutralize the excess acid with dilute sodium hydroxide solution and then perform the tests above which have given positive results with glucose and/or fructose. Repeat for starch but use concentrated hydrochloric acid and boil for 20 minutes.

11 To a suspension of starch in water add a solution of iodine in potassium iodide and water.

12 Optical activity: observe the rotation of the plane of polarization of plane polarized light given by solutions of some carbohydrates in water.

V Organic Chemistry by Qualitative Analysis

Introduction

The preliminary tests should be carried out first. The identification of the elements present should be done with great care because accuracy here is very important: later results may be predictable.

Secondly the homologous series tests should be carried out. If the evidence here is not convincing the tests for another possible series should be carried out and if even that does not clarify the position the identification of the elements present should be repeated.

If the substance is a solid its melting-point should be determined to confirm its identity and to check its purity. If the substance is a liquid it should be converted into a solid derivative for these purposes: the boiling-point determination may lead to decomposition if done at atmospheric pressure.

Typical examples of the members of each homologous series are noted at the start of each table and the analyses set will normally be restricted to these compounds. There are greater dangers inherent in organic than in inorganic qualitative analysis and great care should be exercised: the suggested procedure must not be exceeded. Safety spectacles must be worn at all times.

When writing up the qualitative analysis of organic substances a three column table is the most suitable: test, result and inference. At the end a conclusion as to the nature, if not the identity, of the substance must be drawn.

5.1 Physical Properties

These do not give a conclusive guide.

The physical properties of an organic substance may not give much indication as to its identity. When smelling the substance waft any vapour gently towards you and afterwards breathe out firmly to expel the vapours from your nose.

The substances that are soluble in cold water contain atoms of oxygen and nitrogen which confer polarity, especially if a hydrogen atom is attached to them.

Organic acids in general are weaker acids than those encountered in inorganic analysis. Hydrolysis of compounds is more noticeable and may even be violent.

Test	Result	Inference
1 Physical state	Solid, liquid or gas at room temperature	Do not come to a conclusion that is obviously wrong at the end of your analysis
2 Appearance	Yellow	Iodoform (triiodomethane), 1,3-dinitrobenzene, 2-nitrophenol . . .
	Many other colours	May be due to the cation — see Part III: Inorganic chemistry by qualitative analysis
3 Density	Low	None
	High	Halogen compound or substance contains cation of high relative atomic mass
4 Odour	Pleasant and fruity	Ester
	Pungent	Formic (methanoic) acid, acetic (ethanoic) acid (vinegar), chlorobenzene, acetyl (ethanoyl) chloride, acetic anhydride, benzoyl chloride
	Fishy	Amine
	Disinfectant	Phenol, salicylic (2-hydroxybenzoic) acid
	'TCP'	Halogenated phenol
	Almonds	Benzaldehyde, nitrobenzene
	Ammonia	Ammonium compound
5 Solubility in water	Soluble in cold	Methanol, ethanol, phenol, glycerol (propane-1, 2, 3-triol), aldehydes and ketones of low relative molecular mass, aliphatic acids, ammonium and alkali metal salts of some acids, some amides and amines
	Almost insoluble in cold	Aromatic alcohols and acids, most esters, ethers, hydrocarbons, halogen and nitro compounds
6 pH of aqueous solution	1–4	Acids, esters on hydrolysis, acid chlorides, amine salts
	4–7	Phenols
	7	Alkanes, alkenes, alkynes, many halogen derivatives of alkanes, alcohols, ethers, aldehydes, ketones, esters and nitro compounds
	7–10	Group I and II metals as their salts, aliphatic amines

5.2 The Identification of Elements Present

Test	Result	Inference
1 Heat with excess dry copper(II) oxide in a dry tube (a 1:10 mixture is suitable)	Carbon dioxide evolved: calcium hydroxide test	Carbon
	Steam evolved: anhydrous copper(II) sulphate test	Hydrogen

2 Lassaigne's test

In a hard glass test-tube very carefully fuse a small piece of sodium (half the size of a pea) with the same volume of the substance. Allow the tube to cool slightly before plunging it at arm's length into $10\,cm^3$ distilled water in a boiling tube or mortar and hold a gauze over it. Grind any lumps to a powder and stir well; divide the solution into three portions.

a) Add $1\,cm^3$ iron(II) sulphate solution (if no precipitate add $1\,cm^3$ sodium hydroxide solution in order to obtain a precipitate); boil, cool and add one drop iron(III) chloride solution; then acidify with dilute hydrochloric acid. — Prussian blue precipitate or coloration — Nitrogen

b) Acidify with dilute nitric acid, boil **(in a fume cupboard if test (a) is positive)** then add silver nitrate solution.
- White precipitate — Chlorine
- Cream precipitate — Bromine
- Pale yellow precipitate — Iodine

c) Either acidify with acetic acid then add lead(II) acetate solution — Black precipitate — Sulphur
or add sodium nitroprusside [pentacyanonitrosylferrate(III)] solution. — Purple coloration — Sulphur

---

N.B. There is no positive test for oxygen but if the substance on heating readily yields steam or carbon dioxide **(without burning), then oxygen is probably present.**

1
$$2CuO + (C) \rightarrow 2Cu + CO_2\uparrow$$
$$CO_2 + Ca(OH)_2 \rightarrow CaCO_3\downarrow + H_2O$$
$$CuO + 2(H) \rightarrow Cu + H_2O\uparrow$$
$$\underset{\text{White}}{CuSO_4} + 5H_2O \rightarrow \underset{\text{Blue}}{CuSO_4 \cdot 5H_2O}$$

2
$$Na + (C) + (N) \rightarrow Na^+ + CN^-$$
$$Na + (X) \rightarrow Na^+ + X^-\ (\text{where } X = Cl,\ Br\ or\ I)$$
$$2Na + (S) \rightarrow 2Na^+ + S^{2-}$$

[Zinc and sodium carbonate (2:1 by mass) react similarly in Middleton's version of this test].

$$\underset{\text{Excess}}{2Na + 2H_2O} \rightarrow 2Na^+ + 2OH^- + H_2\uparrow$$

a)
$$Fe^{2+} + 2OH^- \rightarrow \underset{\text{White, then green}}{Fe(OH)_2\downarrow}$$
$$4Fe(OH)_2 + 2H_2O + O_2 \rightarrow 4Fe(OH)_3\downarrow$$
$$Fe^{2+} + 6CN^- \rightarrow [Fe(CN)_6]^{4-}$$
$$Fe(OH)_3 + 3H^+ \rightarrow Fe^{3+} + 3H_2O$$
$$4Fe^{3+} + 3[Fe(CN)_6]^{4-} + 14H_2O \rightarrow \underset{\text{Prussian Blue}}{Fe_4[Fe(CN)_6]_3 \cdot 14H_2O\downarrow}$$

b)
$$H^+ + CN^- \rightarrow HCN\uparrow$$
$$Ag^+ + X^- \rightarrow AgX\downarrow$$

c) $Pb^{2+} + S^{2-} \rightarrow PbS\downarrow$

In the presence of acetic (ethanoic) acid no precipitate of lead hydroxide, cyanide or halide interferes with this test.
$$[Fe(CN)_5NO]^{2-} + S^{2-} \rightarrow [Fe(CN)_5NOS]^{4-}$$

5.3 Heating on a Crucible Lid

Put a small quantity of the substance on a crucible lid and briefly direct the Bunsen flame straight on to it. Consider the following three aspects:

	Result	Inference
1	**a)** Almost non-luminous flame	Aliphatic compound, probably saturated and of low relative molecular mass
	b) Smoky luminous flame	Unsaturated aliphatic or an aromatic compound
	c) Non-flammable	Compound rich in halogen or containing a metal
	d) Violet vapour (iodine)	Iodoform (triiodomethane)
2	**a)** Residual ash	Compound containing a metal – see Part III: Inorganic chemistry by qualitative analysis
	If so, try two further tests now on the ash: Flame test – yellow etc.	Sodium etc.
	Add dilute hydrochloric acid – sulphur dioxide evolved (potassium permanganate [manganate(VII)] solution and dilute sulphuric acid on filter-paper turns from purple to colourless). If no sulphur dioxide evolved residue is probably the metal oxide or carbonate.	Hydrogensulphite addition compound of an aldehyde or ketone
	b) Charring, i.e. substance becomes darker due to formation of carbon. Some acids do not char but their salts do, e.g. succinates (butanedioates) and salicylates (2-hydroxybenzoates)	Most organic substances do this especially carbohydrates, tartaric (dihydroxybutanedioic) and citric (2-hydroxypropane-1,2,3-tricarboxylic) acids and their salts; decomposition of the substances named is usually accompanied by the odour of burnt sugar
3	**a)** Odour of burning wool, hair or flesh	Amino-acid

The luminosity of a flame is due to the excitation of twin particles of carbon. Unsaturated and aromatic substances need a high ratio of air to gas for complete combustion so they frequently yield carbon and steam when burnt.

5.4 Heating with Soda-Lime

Grind together four spatula loads (or 1 cm³) of the compound and four of soda-lime. If there is no reaction in the cold heat the mixture and collect any condensable vapours through a delivery tube into a cold test-tube.

Result	Inference
1 Ammonia evolved (alkaline to damp pH paper, pungent smell)	Ammonium compound or an amide
2 Hydrogen evolved (burns with squeaky 'pop')	Formate (methanoate) or an oxalate (ethanedioate)
3 Hydrocarbon evolved (burns producing steam and carbon dioxide)	Carboxylic acid or a salt
4 Phenol evolved (odour of disinfectant)	Salicylic (2-hydroxybenzoic) acid or a salt; a phenol distils unchanged
5 Odour of burnt sugar	Carbohydrate, tartaric (dihydroxybutanedioic) or citric (2-hydroxypropane-1,2,3-tricarboxylic) acid or one of their salts
6 Amine evolved (alkaline to damp pH paper, fishy smell)	An amine salt, an anilide (phenylamide) or an amino acid

Soda-lime is non-deliquescent: it is composed of sodium hydroxide and calcium hydroxide.

1
$$NH_4^+ + OH^- \rightarrow H_2O + NH_3 \uparrow$$
$$-CONH_2 + OH^- \rightarrow -COO^- + NH_3 \uparrow$$

2
$$HCOO^- + OH^- \rightarrow CO_3^{2-} + H_2 \uparrow$$
$$H_2C_2O_4 + 2OH^- \rightarrow HCOOH + CO_3^{2-} + H_2O$$

3
$$RCOO^- + OH^- \rightarrow CO_3^{2-} + RH \uparrow$$

4

5 No simple equations.

6
$$-NH_3^+X^- + OH^- \rightarrow X^- + H_2O + -NH_2 \uparrow$$
$$RCH(NH_2)COOH + 2OH^- \rightarrow CO_3^{2-} + H_2O + RCH_2NH_2 \uparrow$$

5.5 The Addition of Dilute Sodium Carbonate Solution

Put $4\,cm^3$ of 1M sodium carbonate solution in a test-tube plus an anti-bumping granule, boil gently to ensure absence of hydrogencarbonate ions and of carbon dioxide; cool, add half a spatula load of the compound and again boil gently, passing the gas(es) evolved through calcium hydroxide solution.

Result	Inference
Carbon dioxide evolved (calcium hydroxide solution turns milky)	Acid, ester which hydrolyzes easily [e.g. dimethyl oxalate (ethanedioate)], amine salt, etc

$$2H^+ + CO_3^{2-} \rightarrow H_2O + CO_2\uparrow$$
$$Ca(OH)_2 + CO_2 \rightarrow CaCO_3\downarrow + H_2O$$

5.6 The Addition of Dilute Hydrochloric Acid

Put $3\,cm^3$ of dilute hydrochloric acid in a test-tube and then add half a spatula load of the compound and warm the mixture.

Result	Inference
The substance is soluble in warm acid (but was insoluble in water)	A basic substance, e.g. an aromatic amine or a salt of an aromatic acid

$$RNH_2 + HCl \rightarrow RNH_3^+ + Cl^-$$

5.7 The Addition of Concentrated Sodium Hydroxide Solution

Put 3 cm^3 of concentrated (5M) sodium hydroxide· solution in a test-tube and add half a spatula load of the compound. Note any reaction in the cold, then add an anti-bumping granule and boil the solution for several minutes.

Result	Inference
1 Dissolution easy in alkali (NB almost insoluble in cold water but soluble in hot).	Aromatic carboxylic acid
2 Ammonia evolved in cold	Often indicates an ammonium compound
3 Ammonia evolved on heating	Amide
4 Brown resin	Aliphatic aldehyde (other than formaldehyde) or carbohydrate
5 Slow reaction on heating but no gas evolved	Ester, benzaldehyde, etc.
6 Aromatic amine liberated: an oil rises to the surface	
—rapidly	Amine salt
—slowly	Anilide (phenylamide)

1    $RCOOH + OH^- \rightarrow RCOO^- + H_2O$

2    $NH_4^+ + OH^- \rightarrow H_2O + NH_3 \uparrow$

3    $-NH_2 + OH^- \rightarrow -O^- + NH_3 \uparrow$

5    $RCOOR' + OH^- \rightarrow RCOO^- + R'OH$
     $2C_6H_5CHO + OH^- \rightarrow C_6H_5COO^- + C_6H_5CH_2OH$
         Cannizzaro's reaction

6    $RNH_3^+ X^- + OH^- \rightarrow RNH_2 + H_2O + X^-$
   $RNHCOR' + OH^- \rightarrow RNH_2 + R'COO^-$

5.8 The Addition of Concentrated Sulphuric Acid

Put half a spatula load of the compound in a clean dry test-tube and then add 1 cm^3 of concentrated sulphuric acid. Note any reaction in the cold and then heat the tube, gently at first. When thus heated a large number of organic substances darken to some extent, without going fully black.

Result	Inference
1 No blackening, carbon monoxide evolved (burns without explosion with blue flame, *afterwards* calcium hydroxide solution turns milky)	Formate (methanoate)
2 No blackening, carbon monoxide and dioxide evolved	Oxalate (ethanedioate)
3 No blackening, yellow colour develops, carbon monoxide and dioxide evolved	Citrate (2-hydroxypropane-1,2,3-tricarboxylate)
4 No blackening, substances mix (whereas original substance immiscible with water)	Aromatic hydrocarbon (they form sulphonates), alkene, alkyne
5 No blackening, no marked effervescence, pungent vapour	An acid or its salt, e.g. acetic (ethanoic), succinic (butanedioic), benzoic
6 Blackening with effervescence	Carbohydrate or tartrate (dihydroxybutanedioate)
7 Blackening without effervescence	Polyhydric phenol

1 $HCOO^- \rightarrow OH^- + CO\uparrow$

2 $C_2O_4^{2-} \rightarrow O^{2-} + CO\uparrow + CO_2\uparrow$

4      $RH + H_2SO_4 \rightarrow RSO_3H + H_2O$
      $\rangle C = C\langle + H_2SO_4 \rightarrow \rangle CH - C(SO_3OH)\langle$
      Alkyne similarly

5.9 Tests for Unsaturation

Add half a spatula load of the substance to each of the following reagents.

Reagent	Result	Inference
1 Potassium permanganate [manganate(VII)] solution and dilute sulphuric acid	Goes colourless – if no reaction in the cold warm gently	Unsaturated or oxidizable compound
2 Potassium permanganate solution and sodium carbonate solution (Baeyer's reagent)	Goes green then gives a brown suspension in a colourless solution	Unsaturated compound, primary or secondary alcohol, aldehyde
3 Bromine water	Goes colourless White precipitate	Unsaturated compound Amine or phenol with vacant 2- or 4- positions

Test (2) is better than test (1) for an unsaturated compound because alkaline permanganate is a less powerful oxidizing agent.

1 $2MnO_4^- + 6H^+ + 2H_2O + 5 \rangle C=C\langle \rightarrow 2Mn^{2+} + 5 \rangle COH-COH\langle$
$\qquad\qquad\qquad\qquad\qquad$ Colourless in solution

$5RCH_2OH + 6H^+ + 2MnO_4^- \rightarrow 5RCHO + 2Mn^{2+} + 8H_2O$
A secondary alcohol yields a ketone

$5RCHO + 6H^+ + 2MnO_4^- \rightarrow 5RCOOH + 2Mn^{2+} + 3H_2O$

2 $2MnO_4^- + 2OH^- + \rangle C=C\langle \rightarrow 2MnO_4^{2-} + \rangle COH-COH\langle$
$\qquad\qquad\qquad\qquad\qquad$ Green

$2MnO_4^- + 4H_2O + 3 \rangle C=C\langle \rightarrow 2MnO_2 \downarrow + 2OH^- + 3 \rangle COH-COH\langle$
$\qquad\qquad\qquad\qquad\qquad$ Brown

3 $Br_2 + H_2O \rightleftharpoons HOBr + H^+ + Br^-$

$\rangle C=C\langle + Br_2 \rightarrow \rangle C-C\langle + Br^-$
$\qquad\qquad\qquad\quad \underset{Br^+}{\diagdown\diagup}$
$\qquad\qquad$ bromonium ion

$\underset{Br^+}{\overset{}{\rangle C-C\langle}} + H_2O \rightarrow \rangle C(OH)-CBr\langle + H^+$

Phenol C_6H_5OH reacts likewise (electrophilic substitution).

Test	Result	Inference
1 **Phosphorus pentachloride** To half a spatula load of the substance in a **dry** test-tube add a spatula load of phosphorus pentachloride.	Violent evolution of hydrogen chloride (acidic, pungent smell, white cloud with ammonia)	Alcohol or carboxylic acid. **N.B.** Also water
2 **Neutral iron(III) chloride** To the iron(III) chloride solution add sodium hydroxide solution until a faint precipitate appears; centrifuge and use the solution. If it is the free acid being studied add ammonia solution until the smell is apparent, then in the presence of an anti-bumping granule boil until the smell has gone. Cool. Add a few drops of the iron(III) chloride solution to the substance (in solution if it is not a liquid).	**a)** Red coloration, brown gelatinous precipitate on warming **b)** Yellow coloration **c)** Very faint yellow coloration **d)** Violet coloration – pale deep **e)** Buff or pale brown precipitate; then add dilute hydrochloric acid: – clear solution – white precipitate **f)** If no coloration add dilute hydrochloric acid: green coloration	Formate (methanoate) or acetate (ethanoate) Tartrate (dihydroxy-butanedioate) or citrate (2-hydroxypropane-1, 2, 3-tricarboxylate) Possibly an oxalate (ethanedioate) Phenol or resorcinol (benzene-1,3-diol) Salicylate (2-hydroxybenzoate) Succinate (butanedioate) Benzoate Aniline (phenylamine) or methylaniline (methyl-phenylamine)

1 $-OH + PCl_5 \rightarrow -Cl + POCl_3 + HCl\uparrow$

2 The colorations are given by covalent iron compounds which form in solution: they may be due to chelate compounds.

5.11 Tests for Carbonyl Groups

Test	Result	Inference
1 **Fehling's solution** To a spatula load of the substance add 1 cm^3 sodium carbonate solution and then 2cm^3 Fehling's solution (1 cm^3 each of A and of B if supplied thus). Boil gently for one minute.	The mixture turns green then yellow and finally leaves a red precipitate	Aldehyde, formate (methanoate), chloroform (trichloromethane), iodoform (triiodomethane), reducing sugar or possibly an ester or a phenol
2 **Tollen's reagent (silver mirror test)** Put 3 cm^3 silver nitrate solution in a thoroughly clean test-tube, add three drops sodium hydroxide solution, then ammonia solution dropwise until the precipitate has just redissolved; finally add three drops of the substance and warm gently, with shaking.	Silver mirror or precipitate	Aldehyde, formate, tartrate (dihydroxybutanedioate), reducing sugar or possibly an amine
3 Schiff's reagent and 4 2, 4-dinitrophenylhydrazine (in 3M hydrochloric acid, Brady's reagent)	These tests, which could be applied here if the nature of the compound is almost decided, are described in section 5.16 parts 9 and 10	

1 $-CHO + 2Cu^{2+} + 5OH^- \rightarrow Cu_2O\downarrow + -COO^- + 3H_2O$

2 $-CHO + 2\,[Ag(NH_3)_2]^+ + H_2O \rightarrow 2\,Ag\downarrow + -COO^- + 3NH_4^+ + NH_3$

5.12 Hydrocarbons

Hydrocarbons, e.g. hexane, cyclohexane, benzene and toluene (methylbenzene) are colourless liquids, insoluble in and less dense than water. The tests in section 5.9 should have given a negative result. Tests (2) and (3) below are best restricted to the aromatic hydrocarbons.

---

1 **Boiling-point**

Hexane	342K	(69°C)
Cyclohexane	354K	(81°C)
Benzene	353K	(80°C); vapour poisonous
Toluene	384K	(111°C)

---

2 **Nitration**

To 1 cm^3 concentrated nitric acid add 1 cm^3 concentrated (or fuming) sulphuric acid and then carefully add 0.5 cm^3 of the substance: cool the mixture if the reaction becomes too violent. Shake the mixture for two minutes and then pour it into about 10 cm^3 of cold water: the nitro compound separates as a dense yellow oil or solid.

$$-H + NO_2^+ \rightarrow -NO_2 + H^+$$

---

3 **Sulphonation**

To 1 cm^3 of the substance add slowly 1 cm^3 of concentrated sulphuric acid and shake carefully: a homogeneous solution should be obtained. Pour the mixture into about 10 cm^3 of cold water, stirring all the time: a clear solution of the sulphonic acid is obtained. **N.B.** The original substance is immiscible with water.

$$-H + H_2SO_4 \rightarrow -SO_3H + H_2O$$

5.13 Halogen Compounds

Iodomethane, iodoform (triiodomethane), iodoethane, bromoethane, chloroform (trichloromethane), carbon tetrachloride (tetrachloromethane), chlorobenzene, benzyl chloride [(chloromethyl) benzene] may be studied. These halogen compounds are all liquids except iodoform which is a yellow solid. They are denser than and insoluble in water. Chlorobenzene has an aromatic odour, (chloromethyl)benzene a pungent odour.

Test	Result	Inference
1 **Cold silver nitrate solution** Add a few drops of the substance to 2 cm^3 of silver nitrate solution and shake vigorously for several minutes	A definite precipitate Faint precipitate in the cold To the precipitate add **a)** dilute nitric acid and boil **b)** ammonia solution	Iodomethane, iodo-ethane or bromoethane (Chloromethyl)benzene **a)** Insoluble if a halide. **b)** soluble if a chloride; partially soluble if a bromide; insoluble if an iodide
2 **Alkaline hydrolysis** Put 1 cm^3 of the substance and an anti-bumping granule with 10 cm^3 of sodium hydroxide solution in ethanol in a small flask fitted with a reflux condenser; heat for ten minutes. If a white solid has separated dilute the solution with water. Acidify the solution with dilute nitric acid and then add silver nitrate solution.	White precipitate Cream precipitate Pale yellow precipitate	Chloride Bromide Iodide
3 **Other tests** (i) Fehling's solution reduced on heating (ii) Heating the substance	see section 5.11 Iodine released – see section 5.3	Chloroform Iodoform

1 $-X + H_2O \rightarrow -OH + H^+ + X^-$
   (X = Cl, Br or I)
   $Ag^+ + X^- \rightarrow AgX\downarrow$
   $AgX + 2NH_3 \rightleftharpoons [Ag(NH_3)_2]^+ + X^-$

2 $-X + OH^- \rightarrow -OH + X^-$
   Chlorobenzene does not hydrolyze under these conditions and so no proof is given.

5.14 Alcohols

Alcohols, e.g. methanol, ethanol and propan-1(or 2)-ol are colourless liquids, completely miscible with water and have a faint odour; glycerol (propane-1,2,3-triol) is viscous, miscible with water and odourless.

---

1 Methanol

a) Heat 1 cm^3 of the substance with a spatula load of salicylic (2-hydroxybenzoic) acid and a few drops of concentrated sulphuric acid for one minute. Cool and pour the mixture into about 15 cm^3 of water and note the smell of methyl salicylate (oil of wintergreen).

b) The iodoform test described in (2b) below is negative.

c) Put five spatula loads of potassium dichromate [dichromate(VI)] and 10 cm^3 of dilute sulphuric acid in a flask then carefully add with constant swirling about 3 cm^3 of concentrated sulphuric acid; finally add 1 cm^3 of the substance and an anti-bumping granule. Boil the mixture under a reflux condenser for five minutes then distil off some of the liquid and test it for formic (methanoic) acid as in section 5.18. Most of the formic acid may have been oxidised to carbon dioxide.

$$2Cr_2O_7^{2-} + 16H^+ + 3CH_3OH \rightarrow 4Cr^{3+} + 3HCOOH + 11H_2O$$
Orange   Green

---

2 Ethanol

a) Ethyl salicylate, which can be made in an experiment similar to (1a) above, has a similar odour to methyl salicylate but the odour of the methyl derivative is more characteristic. In a test-tube put 1 cm^3 of the substance and 1 cm^3 of glacial acetic acid finally adding a few drops of concentrated sulphuric acid; heat the tube for one minute and then pour the mixture into about 20 cm^3 of water: the fruity odour of ethyl acetate should be apparent.

$$C_2H_5OH + CH_3COOH \rightleftharpoons CH_3COOC_2H_5 + H_2O$$
(Equilibrium constant = 4)

b) The iodoform test: to 0.2 cm^3 of the substance add 2 cm^3 0.5M potassium iodide solution and 4 cm^3 sodium hypochlorite [chlorate(I)] solution. Warm the mixture to 323 K (50°C) for two minutes and then cool it: fine yellow crystals separate.

$$OCl^- + I^- \rightarrow OI^- + Cl^-$$
$$C_2H_5OH + OI^- \rightarrow CH_3CHO + H_2O + I^-$$
$$CH_3CHO + 3OI^- \rightarrow CI_3CHO + 3OH^-$$
$$CI_3CHO + OH^- \rightarrow CHI_3\downarrow + HCOO^-$$

Ethanol, propan-2-ol, acetaldehyde (ethanal) and acetone (propanone) are common substances that give a positive iodoform test, i.e. the group CH_3CO- must either be present originally or be formed by oxidation of the parent substance. Alternative reagents are 2 cm^3 iodine solution just decolorized by the addition of sodium hydroxide.

c) Make up 4 cm^3 of a concentrated solution of potassium dichromate [dichromate(VI)] and add 1 cm^3 of concentrated sulphuric acid, warm the mixture gently and then add 0.5 cm^3 of the substance: the solution turns green and smells of acetaldehyde. After a few moments the sharp smell of acetic acid may be noticed.

$$Cr_2O_7^{2-} + 8H^+ + 3C_2H_5OH \rightarrow 2Cr^{3+} + 3CH_3CHO + 7H_2O$$
$$Cr_2O_7^{2-} + 8H^+ + 3CH_3CHO \rightarrow 2Cr^{3+} + 3CH_3COOH + 4H_2O$$

---

3 Glycerol

In a fume cupboard: heat 0.5 cm^3 of the substance with two spatula loads of finely powdered potassium hydrogensulphate. Smell the mixture cautiously: it has the characteristic irritative odour of acrolein (propenal).

5.15 Phenols

Phenols, e.g. phenol and resorcinol (benzene-1,3-diol), are colourless solids when pure, soluble in water and in sodium hydroxide solution. See also section 5.9(3).

Care: phenol is very corrosive.

Test	Result and Inference	
	Phenol	*Resorcinol*
1 Liebermann's reaction Put a few crystals of sodium nitrite [nitrate(III)] and one spatula load of the substance in a clean dry test-tube and then heat it for 20 seconds. Cool and add twice the volume of concentrated sulphuric acid, stir, pause a few minutes:		
a) initial colour	Green or blue	Green or blue
b) then add water carefully	Red	Red
c) finally add sodium hydroxide solution	Green or blue	Brown

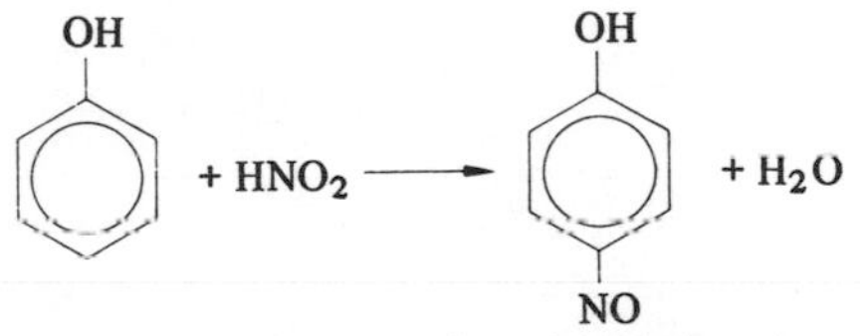

4 — nitrosophenol

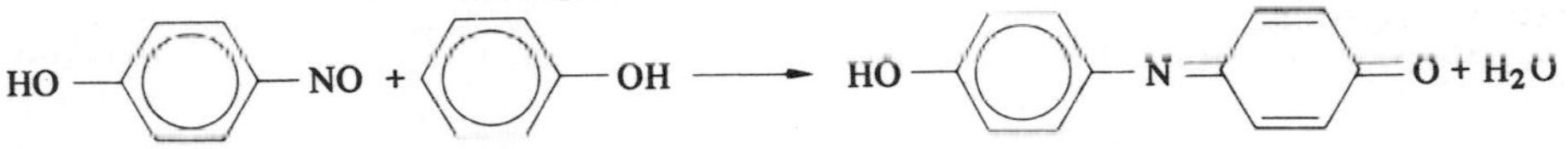

(resorcinol reacts likewise)

an indophenol

2 Azo dye formation Dissolve three drops of aniline (phenylamine) in 1 cm^3 of concentrated hydrochloric acid and 3 cm^3 of water. Shake, cool in ice (if possible) and add a few drops of sodium nitrite solution. Add this cold solution to a cold solution of the substance in excess sodium hydroxide solution	Orange	Red

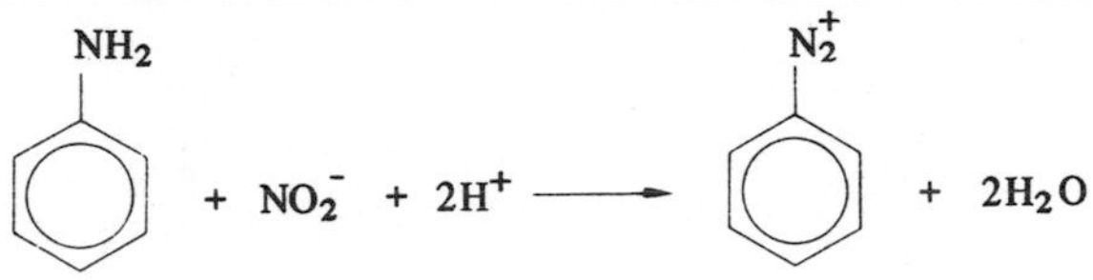

(resorcinol reacts likewise)

3 Phthalein reaction

Put one spatula load of the substance and one spatula load of phthalic (benzene-1, 2-dicarboxylic) anhydride in a dry test-tube. Add two drops of concentrated sulphuric acid and warm gently for one minute with great care. Cool and add dilute sodium hydroxide solution in excess.

Red
(phenolphthalein)

Fluorescent green
(fluorescein)

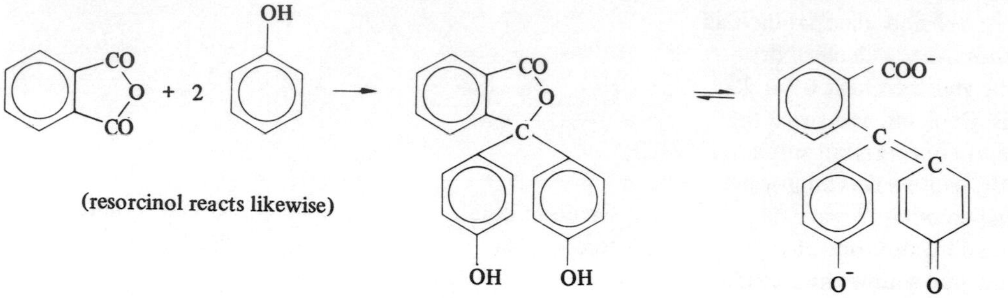

(resorcinol reacts likewise)

phenolphthalein (colourless form)

phenolphthalein (red form)

5.16 Aldehydes

Common aldehydes are formaldehyde (methanal), usually encountered as a 40% solution in water (formalin) rather than as a gas — pungent odour; acetaldehyde (ethanal), b.p. 294 K (21 °C), miscible with water — acrid odour and benzaldehyde, b.p. 452 K (179 °C), insoluble in water — odour of bitter almonds.

Test	Result and Inference		
	Formaldehyde	*Acetaldehyde*	*Benzaldehyde*
1 Fehling's solution — see section 5.11(1)	Red precipitate	Red precipitate	Red precipitate very slowly forms
2 Tollen's reagent — see section 5.11(2)	Silver mirror	Silver mirror	Silver mirror
3 To 0.5 cm³ of the substance add 3 cm³ of potassium dichromate [dichromate (VI)] solution and 2 cm³ of dilute sulphuric acid. Warm the mixture gently.	Turns green	Turns green	Turns green
4 To 0.5 cm³ of the substance add 3 cm³ of potassium permanganate [manganate (VII)] solution and 2 cm³ of dilute sulphuric acid. Warm the mixture gently.	Turns colourless	Turns colourless	Turns colourless
5 Add 3 cm³ of saturated sodium hydrogensulphite solution to 1 cm³ of the compound.	— (Other aliphatic aldehydes may give a white precipitate)	—	White precipitate
6 Warm 1 cm³ of the substance with 1 cm³ of concentrated sodium hydroxide solution, cool then add 2 cm³ concentrated hydrochloric acid.	—	Yellow resin, odour of bad apples	White precipitate of benzoic acid on cooling acidic solution
7 To 0.5 cm³ of the substance add 1 cm³ fresh solution of sodium nitroprusside [pentacyanonitrosylferrate(III)] then add excess sodium hydroxide solution.	—	Red coloration	—
8 The iodoform test — see section 5.14(2b)	—	Yellow crystals	—
9 The addition of 1 cm³ of Schiff's reagent to 0.5 cm³ of the cold compound.	Red coloration	Red coloration	Slowly turns red

10 To 0.5 cm³ of the substance add 2 cm³ of 2,4-dinitrophenylhydrazine solution (in 3M hydrochloric acid, Brady's reagent). Warm if no reaction in cold, then cool.

	Yellow precipitate in cold	Yellow-orange precipitate	Red precipitate

1 $-CHO + 2Cu^{2+} + 5OH^- \rightarrow Cu_2O\downarrow + -COO^- + 3H_2O$

2 $-CHO + 2\,[Ag(NH_3)_2]^+ + H_2O \rightarrow 2\,Ag\downarrow + -COO^- + 3NH_4^+ + NH_3$

3 $3RCHO + 8H^+ + Cr_2O_7^{2-} \rightarrow 2Cr^{3+} + 3RCOOH + 4H_2O$

4 $5RCHO + 6H^+ + 2MnO_4^- \rightarrow 2Mn^{2+} + 5RCOOH + 3H_2O$

5 $C_6H_5CHO + Na^+ + HSO_3^- \rightarrow C_6H_5-\overset{\displaystyle H}{\underset{\displaystyle OH}{C}}-OSO_2^-\,Na^+\downarrow$

6 The yellow resin is possibly a linear polymer $CH_3(CH=CH)_n CHO$. Formaldehyde and benzaldehyde undergo the disproportionation in Cannizzaro's reaction but only the latter substance gives a visible indication.

$2C_6H_5CHO + OH^- \rightarrow C_6H_5COO^- + C_6H_5CH_2OH$

$C_6H_5COO^- + H^+ \rightarrow C_6H_5COOH\downarrow$

10 $>C=O + H_2NNH-\!\!\!\!\raisebox{-0.3em}{\text{(ring, NO}_2\text{)}}\!\!\!-NO_2 \longrightarrow >C=NNH-\!\!\!\!\raisebox{-0.3em}{\text{(ring, NO}_2\text{)}}\!\!\!-NO_2 + H_2O$

152

152

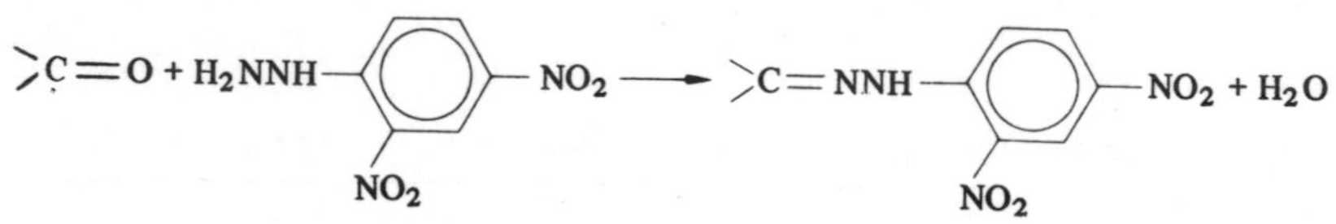

5.17 Ketones

Common ketones are acetone (propanone) b.p. 330 K (57°C) — characteristic odour; ethyl methyl ketone (butanone), b.p. 353 K (80°C) and acetophenone (phenylethanone, methyl phenyl ketone), b.p. 475 K (202°C). The results for ketones should be compared with those for aldehydes (see section 5.16).

Test	Result
1 Fehling's solution — see section 5.11(1)	Negative
2 Tollen's reagent — see section 5.11(2)	Negative
3 Potassium dichromate — see section 5.16(3)	Negative
4 Potassium permanganate — see section 5.16(4)	Negative
5 Sodium hydrogensulphite — see section 5.16(5)	White precipitate given by simple ketones
6 Sodium nitroprusside — see section 5.16(7)	Red coloration
7 The iodoform test — see section 5.14(2b)	Positive (yellow crystals), if a methyl ketone
8 Schiff's reagent — see section 5.16(9)	Acetone causes a red coloration; other ketones may do so slowly, if at all
9 Brady's reagent see section 5.16(10)	Positive (orange precipitate)
10 To 1 cm^3 of the substance add half a spatula load of finely powdered 1,3-dinitrobenzene and then excess sodium hydroxide solution; shake.	Violet coloration which slowly fades

5.18 Carboxylic Acids

Formic (methanoic), acetic (ethanoic), oxalic (ethane-dioic), succinic (butanedioic), tartaric (dihydroxy-butanedioic), citric (2-hydroxypropane-1,2,3-tricarboxy-lic), benzoic and salicylic (2-hydroxybenzoic) acids are common carboxylic acids. Only the first two are liquids at room temperature; the aliphatic ones (the first six) are soluble in cold water and the aromatic ones are sparingly soluble in cold water but soluble in hot water.

Test	Result and Inference		
	Formic	*Acetic*	*Oxalic*
1 Heating with soda-lime — see section 5.4	Hydrogen ignites explosively	Methane ignites smoothly	Hydrogen ignites explosively
2 The addition of concentrated sulphuric acid — see section 5.8	No charring but carbon monoxide produced	No charring but pungent smell	No charring but carbon monoxide and dioxide produced
3 Baeyer's reagent — see section 5.9	Decolorized	—	Slowly decolorized, only rapid above 340K (67°C)
4 Neutral iron(III) chloride solution — see section 5.10	Red coloration, brown gelatinous precipitate on boiling	Red coloration, brown gelatinous precipitate on boiling	Possibly a faint yellow coloration
5 Tollen's reagent — see section 5.11	Silver mirror	—	—
6 Add calcium chloride solution to a neutralized (see section 5.10) solution of the acid.	—	—	White precipitate, insoluble in acetic acid
7 The fluorescein test [similar to the phthalein reaction — see section 5.15(3)] : use one spatula load of the substance and one of resorcinol (benzene-1,3-diol).	—	—	—
8 Fenton's reagent: to 2 cm³ of a solution of the compound add one drop of fresh iron(II) sulphate solution and two drops of hydrogen peroxide followed by excess sodium hydroxide solution	—	—	—

Succinic	Tartaric	Citric	Benzoic	Salicylic
Ethane	Odour of burnt sugar	Odour of burnt sugar	Benzene	Phenol
No charring but pungent smell	Charring and carbon monoxide, carbon dioxide and sulphur dioxide produced	Goes yellow and carbon monoxide and dioxide produced	No charring but pungent smell	Slight charring
—	Rapidly decolorized on heating	Rapidly decolorized on heating	—	Decolorized
Buff precipitate, soluble in dilute hydrochloric acid	Yellow coloration	Yellow coloration	Buff precipitate, with dilute hydrochloric acid turns white	Violet coloration
—	Silver mirror	—	—	—
White precipitate on boiling, soluble in acetic acid	White precipitate, soluble in acetic acid	White precipitate on boiling, insoluble in acetic acid	—	—
Red solution with bright green fluorescence	—	—	—	—
—	Violet coloration, intensified by adding one drop of iron(III) chloride solution	—	—	—

5.19 Esters

Common esters are the methyl and ethyl esters of formic (methanoic), acetic (ethanoic), oxalic (ethane-dioic), benzoic and salicylic (2-hydroxybenzoic) acids. Of these only dimethyl oxalate is a solid; methyl formate and oxalate are soluble in water but the others are only sparingly soluble.

The esters have to be identified by hydrolysis: put $2 \, cm^3$ of the substance in a flask, add $20 \, cm^3$ of sodium hydroxide solution and a few anti-bumping granules: boil under a reflux condenser for 15 minutes, cool and rearrange for distillation. Obtain $5 \, cm^3$ of distillate.

1 Test the distillate for an alcohol as in section 5.14, remembering water is present. Alternatively drip the distillate into five spatula loads of potassium dichromate [dichromate(VI)] and $10 \, cm^3$ of dilute sulphuric acid etc. as in 5.14(1c); then identify the aldehyde or ketone as in 5.16(2), (4) and (10).

2 Cool the mixture in the flask and acidify it with dilute sulphuric acid.
   a) If a precipitate forms the acid is aromatic: test it as in sections 5.18(1), (2) and (4).
   b) If no precipitate is formed the acid is aliphatic: it may be volatile (e.g. formic or acetic) or non-volatile (e.g. oxalic). Distil off a further $5 \, cm^3$ and/or test the solution in the flask as in sections 5.18(3) to (8).

$$\begin{array}{l} R \\ {\scriptstyle\diagdown}C{=}O + OH^- \;\rightarrow\; \\ R'O \end{array} \begin{array}{l} R \\ {\scriptstyle\diagdown}C{=}O + R'OH \\ {}^-O \end{array}$$

$$RCOO^- + H^+ \;\rightarrow\; RCOOH$$

5.20 Ammonium Salts and Amides

The most likely substances to be encountered are the ammonium salts of the acids in section 5.18; acetamide (ethanamide), oxamide (ethanediamide), urea (carb-amide) and benzamide. The ammonium salts are readily soluble in water, acetamide and urea are also soluble.

Reactions of ammonium salts

1 They may give off ammonia when sodium hydroxide solution is added in the cold.

2 Their aqueous solutions give the characteristic reactions of the acid in a neutralized solution — see section 5.18.

Reactions of amides

1 Ammonia is evolved on heating with sodium hydroxide solution.

2 When aqueous solutions are obtainable they do not react with iron(III) chloride.

3 The parent acid should be identified as follows: boil four spatula loads of the substance and $20 \, cm^3$ of sodium hydroxide solution together with an anti-bumping granule under a reflux condenser for 15 minutes, cool the flask, acidify the contents with dilute sulphuric acid and cool again.
   a) If a precipitate forms the acid is aromatic: test it as in sections 5.18 (1), (2) and (4).
   b) The rapid evolution of carbon dioxide without any precipitation indicates urea.
   $$CO(NH_2)_2 + 2OH^- \;\rightarrow\; CO_3^{2-} + 2NH_3 \uparrow$$
   $$2H^+ + CO_3^{2-} \;\rightarrow\; H_2O + CO_2 \uparrow$$
   This should be confirmed by the biuret test: heat half a spatula load of the substance gently to its melting point — ammonia is released and the liquid resolidifies.
   $$2CO(NH_2)_2 \;\rightarrow\; H_2NCONHCONH_2 + NH_3 \uparrow$$
   Dissolve the residue in warm sodium hydroxide solution, cool and add one drop of a very dilute solution (about 0.01 M) of copper(II) sulphate: a purple coloration is given if the substance was urea.
   c) If no precipitate is formed and there is no rapid effervescence the acid is aliphatic: it may be volatile (e.g. acetic) or non-volatile (e.g. oxalic). Distil off a further $5 \, cm^3$ and/or test the solution in the flask as in sections 5.18 (3)–(8).

5.21 Amino Acids

Glycine (aminoethanoic acid) — colourless crystals which are soluble in water — is a typical amino-acid.

1 Glycine is soluble in hot sodium carbonate solution and carbon dioxide is evolved very very slowly.
$$2H_2NCH_2COOH + CO_3^{2-} \rightarrow 2H_2NCH_2COO^- + H_2O + CO_2\uparrow$$

2 Sørensen's reaction: dissolve half a spatula load of the substance in $5\,cm^3$ of water, add two drops of phenolphthalein and then very dilute sodium hydroxide solution dropwise until the solution just turns pink. In a second test-tube put $2\,cm^3$ of formalin (40% formaldehyde, 60% water), add two drops of phenolphthalein and then very dilute sodium hydroxide solution dropwise until the solution just turns pink. Pour the contents of the second tube into the first and, if glycine is the substance, the solutions are immediately decolorized.

3 Dissolve one spatula load of copper(II) carbonate in dilute acetic (ethanoic) acid and then add the substance: glycine intensifies the blue colour.

4 Heat the substance with soda-lime: glycine yields methylamine which has a fishy smell and is alkaline to damp pH paper.
$$H_2NCH_2COOH + 2OH^- \rightarrow CO_3^{2-} + H_2O + CH_3NH_2\uparrow$$

5 To one spatula load of sodium nitrite [nitrate(III)] add $2\,cm^3$ dilute acetic acid, then, because nitrous [nitric(III)] acid is very unstable, cool the mixture before adding half a spatula load of the substance: glycine yields nitrogen.
$$H_2NCH_2COOH + HNO_2 \rightarrow HOCH_2COOH + H_2O + N_2\uparrow$$

5.22 Nitro Compounds

Two common examples are nitrobenzene (pale yellow liquid, insoluble in water, smell of bitter almonds) and 1,3-dinitrobenzene (colourless or pale yellow solid, insoluble in water).

Nitrobenzene

N.B. The nitration test employed for confirmation must not be carried out if the original substance is soluble in water.

In a boiling tube mix $2\,cm^3$ of concentrated nitric acid and $2\,cm^3$ of concentrated sulphuric acid: this is the usual nitration mixture. Add $1\,cm^3$ of the substance with shaking and heat the mixture in a fume cupboard with constant oscillation. Pause a minute then pour the mixture into about $30\,cm^3$ of cold water in a beaker and note the separation of the solid product: 1,3-dinitrobenzene.

1,3-Dinitrobenzene

Dissolve a few crystals of the substance in $2\,cm^3$ of acetone and add a few drops of sodium hydroxide solution. A deep violet coloration is produced which is turned red by acetic acid but destroyed by mineral acids (e.g. hydrochloric).

5.23 Amines

Aniline (phenylamine), 4-methylaniline [4-methylphenyl-amine], a solid, m.p. 318K (45 °C), N-mono- and N,N-dimethylaniline (N-methylphenylamine and N,N-dimethylphenylamine) are representative examples of aromatic amines. They are colourless when pure but soon darken in air. They are soluble in dilute acids; the first two are sparingly soluble in water and the second two insoluble. N-methylaniline is often contaminated with aniline.

Test	Result and Inference			
	Aniline	*4-methyl-aniline*	*N-methyl-aniline*	*N,N-dimethyl-aniline*
1 Isocyanide test: done in a fume cupboard. To 0.25 cm^3 (or half a spatula load) of the substance add three drops of chloroform and 3 cm^3 of a solution of sodium hydroxide in ethanol. Warm the mixture and then destroy the product by pouring it into concentrated hydrochloric acid (rinse the test-tube with acid also).	Foul smell	Foul smell	—	—
2 Diazotization: dissolve 0.25 cm^3 of the substance in 1 cm^3 of concentrated hydrochloric acid, then add 3 cm^3 of water, cool (with ice, if possible) and then add a few drops of a cold solution of sodium nitrite [nitrate(III)].	—	—	Yellow oil precipitated	—
a) If a red colour develops add sodium hydroxide solution.	—	—	—	Green precipitate
b) If there is no red coloration add the cold mixture to a cold solution of resorcinol dissolved in excess sodium hydroxide solution.	Red dye	Red dye	—	—
3 Shake one drop (or crystal) of the compound with 5 cm^3 of water and then add two drops of sodium hypochlorite [chlorate(I)] solution.	Purple then brown coloration	Yellow coloration	Purple then brown coloration	—
4 Dissolve a drop (or crystal) of the substance in a few drops of dilute hydrochloric acid then add three drops of iron(III) chloride solution.	Pale green coloration develops slowly	–	Green coloration	—

1 $C_6H_5NH_2 + CHCl_3 + 3OH^- \rightarrow C_6H_5NC + 3Cl^- + 3H_2O$ (or $C_6H_5NHCH_3$)

2 For azo dye formation reactions see section 5.15(2): here resorcinol is used instead of phenol, coupling takes place in position four.

5.24 Anilides

Two common anilides are acetanilide (*N*-phenylethanamide, sparingly soluble in cold water but soluble in hot) and benzanilide (*N*-phenylbenzamide, sparingly soluble in cold or hot water but soluble in ethanol). They are colourless, odourless, crystalline solids.

1 The isocyanide test is given because they hydrolyze
   — see section 5.23(1).

2 The two anilides mentioned are distinguished by hydrolysis: add $10 \, cm^3$ of concentrated sulphuric acid to $5 \, cm^3$ of water containing an anti-bumping granule and then add two spatula loads of the substance and boil the mixture under a reflux condenser for 15 minutes. Cool the flask.
   a) If a precipitate forms this may mean that the substance was benzanilide which, under these conditions, has yielded benzoic acid: test it as in section 5.18(1), (2) and (4).
   b) If no crystals appear this may mean that the substance was acetanilide which, under these conditions, has yielded acetic (ethanoic) acid: distil off a sample and test it as in section 5.18(4).

5.25 Carbohydrates

Four representatives of a large class of compounds are glucose, fructose, sucrose and starch. They are colourless solids, soluble in water (starch only partially so), insoluble in diethyl ether (ethoxyethane) and unstable to heat.

Test	Result and Inference		
	Glucose	*Fructose*	*Sucrose*
1 Molisch's test: dissolve half a spatula load of the substance in 3 cm^3 of water, add three drops of 1-naphthol in water and ethanol (4:1) then carefully pour in 2 cm^3 of concentrated sulphuric acid so that it forms a lower layer.	Violet coloration	Violet coloration	Violet coloration
2 Warm half a spatula load of the substance with 1 cm^3 of concentrated sulphuric acid using a small flame.	Immediate charring; on further heating carbon dioxide, carbon monoxide and sulphur dioxide evolved		
3 Boil half a spatula load of the substance with 4 cm^3 of sodium hydroxide solution.	Yellow to brown resin with odour of caramel		—
4 Fehling's solution – see section 5.11(1)	Red precipitate	Red precipitate	—
5 Tollen's reagent – see section 5.11(2)	Silver mirror	Silver mirror	—
6 Rapid furfural test: boil together 0.5 cm^3 of a dilute solution of the substance, 0.5 cm^3 of a 1% solution of naphth-1-ol in ethanol and 4 cm^3 of concentrated hydrochloric acid.	Violet coloration: after a few minutes	Violet coloration: immediately	Violet coloration: immediately
7 Osazone formation: dissolve half a spatula load of the substance in 2 cm^3 of water, add five drops of 2,4-dinitrophenylhydrazine solution [in 3M hydrochloric acid (Brady's reagent)] and five drops of glacial acetic acid, warm in a beaker of water.	Yellow crystals: in 15 minutes	Yellow crystals: in 5 minutes	—

8 The test for starch is to add a solution of iodine in potassium iodide and water: a blue coloration is produced due to the formation of a tunnel clathrate.

5.26 Derivatives for Melting-point Determination

See the Introduction (page 135) to this Part of the book and also experiments 4.1 and 4.32.

5.27–5.37

Unknowns which are wholly or partially organic may be set for analysis by a series of tests, no others being permitted. It may be possible in some cases to deduce the precise identity of the substance but in other cases only the nature and possible identity can be deduced. All the practical observations should be recorded and the best possible conclusion sought.

5.27 Substance A

1 Moisten some dry copper(II) oxide with a few drops of A and warm the mixture. Identify the two gases evolved.

2 Mix 0.5 cm³ of substance A with one spatula load of anhydrous sodium carbonate and two of zinc dust. Heat the mixture to bright red heat and maintain it at that temperature for two minutes. Allow the tube and contents to cool before adding water to the residue. Test the solution for the presence of a halide and the residue for a sulphide.

3 To 0.2 cm³ of substance A add 2 cm³ 0.5M potassium iodide solution and 4 cm³ sodium hypochlorite [chlorate(I)] solution. Warm the mixture to 323K (50°C) for two minutes and then cool it.

4 In a fume cupboard put 0.2 cm³ of substance A in a dry test-tube and then add a few grains of phosphorus pentachloride.

5 Make up 4 cm³ of a concentrated solution of potassium dichromate [dichromate(VI)] and add 1 cm³ of concentrated sulphuric acid. Then add 0.5 cm³ of substance A and warm the mixture gently.

5.28 Substance B

1 Put a few drops of substance B on a crucible lid and ignite them.

2 To 0.2 cm³ of substance B add 2 cm³ 0.5M potassium iodide solution and 4 cm³ sodium hypochlorite [chlorate(I)] solution. Warm the mixture to 323K (50°C) for two minutes and then cool it.

3 To 0.5 cm³ of substance B add 1 cm³ 1M sodium carbonate solution and then 2 cm³ Fehling's solution (1 cm³ of A and 1 cm³ of B). Boil gently for one minute.

4 To 0.5 cm³ of substance B add 3 cm³ of potassium dichromate [dichromate(VI)] solution and a few drops of concentrated sulphuric acid. Warm the mixture gently.

5 To 0.5 cm³ of substance B add 2 cm³ of 2,4-dinitrophenylhydrazine solution. Warm if no reaction in the cold, then cool.

5.29 Substance C

1 Moisten some dry copper(II) oxide with a few drops of substance C and warm the mixture. Identify the two gases evolved.

2 To 0.2 cm³ of substance C add 2 cm³ 0.5M potassium iodide solution and 4 cm³ sodium hypochlorite [chlorate(I)] solution. Warm the mixture to 323K (50°C) for two minutes and then cool it.

3 In a fume cupboard put 0.2 cm³ of substance C in a dry test-tube and then add a few grains of phosphorus pentachloride.

4 Warm a mixture of 0.5 cm³ of substance C with 1 cm³ potassium dichromate [dichromate(VI)] solution and 1 cm³ of dilute sulphuric acid.

5 To 0.5 cm³ of substance C add 1 cm³ 1M sodium carbonate solution and then 2 cm³ Fehling's solution (1 cm³ of A and 1 cm³ of B). Boil gently for one minute.

5.30 Substance D

1 Put a few crystals of substance D on a crucible lid and ignite them.

2 Assess the relative solubility of substance D in **(a)** water, **(b)** dilute hydrochloric acid, **(c)** dilute sodium hydroxide solution.

3 Dissolve a spatula load of substance D in about 9 cm^3 of the most suitable solvent above and then divide the solution into three portions for the following tests.
   a) Add a slight excess of dilute hydrochloric acid and then heat the solution to boiling-point before allowing it to cool.
   b) Add bromine water dropwise.
   c) Add potassium permanganate [manganate(VII)] solution dropwise.

5.31 Substance E

1 Heat one spatula load of substance E and then allow it to cool before adding dilute hydrochloric acid to the residue. Identify any gas given off at each stage.

2 To 1 cm^3 of concentrated sulphuric acid add half a spatula load of substance E and warm the mixture gently. Identify any gas evolved.

3 Do a flame test on substance E.

4 Put 3 cm^3 of silver nitrate solution in a thoroughly clean test-tube, add three drops of sodium hydroxide solution, then ammonia solution dropwise until the precipitate has just redissolved; finally sprinkle in a few crystals of substance E and warm the mixture gently.

5 To 2 cm^3 of a solution of substance E in water add 2 cm^3 of mercury(II) chloride solution and warm the mixture gently.

6 To 2 cm^3 of a solution of substance E in water add 2 cm^3 of ammonium carbonate solution.

5.32 Substance F

1 Perform a flame test using some crystals of substance F.

2 Dissolve a few crystals of substance F in 2 cm^3 of water and then add 1 cm^3 of calcium chloride solution. Note the appearance of the contents of the test-tube and then add 1 cm^3 of dilute acetic (ethanoic) acid followed by 2 cm^3 of dilute hydrochloric acid.

3 Put one spatula load of substance F in a test-tube and add 2 cm^3 of concentrated sulphuric acid. Carefully warm the mixture and identify the gas(es) evolved.

4 Dissolve a few crystals of substance F in 2 cm^3 of water and then add 2 cm^3 of dilute sulphuric acid. Warm the test-tube and then add potassium permanganate [manganate(VII)] solution dropwise.

5.33 Substance G

1 Heat some crystals of substance G on a crucible lid, gently at first and finally to red heat. When the residue is cold add dilute hydrochloric acid.

2 Put one spatula load of substance G in a test-tube and add 3 cm^3 of dilute sulphuric acid. Bring the mixture to the boil and then pour it into 15 cm^3 of water in a beaker and smell the vapour.

3 Put one spatula load of substance G in a test-tube, add 3 cm^3 of ethanol and then, carefully, 1 cm^3 of concentrated sulphuric acid. Note the enthalpy of mixing and then warm the tube carefully for a few minutes. Pour the mixture into a beaker containing about 15 cm^3 of water and smell the vapour.

4 Dissolve a few crystals of substance G in 2 cm^3 of water and then add 1 cm^3 of iron(III) chloride solution. Boil the mixture and then add dilute hydrochloric acid.

5.34 Substance H

1 Mix a spatula load of substance H with an equal quantity of soda-lime and heat the mixture in a test-tube.

2 Dissolve a few crystals of substance H in $2\,cm^3$ of water, acidify with dilute hydrochloric acid and heat the mixture until it boils; then allow it to cool.

3 Dissolve a few crystals of substance H in $2\,cm^3$ of water and add first $1\,cm^3$ of iron(III) chloride solution and secondly $1\,cm^3$ of dilute hydrochloric acid.

4 To a spatula load of substance H add $2\,cm^3$ of concentrated sulphuric acid and carefully warm the mixture.

5 To a spatula load of substance H add $2\,cm^3$ of methanol and then a few drops of concentrated sulphuric acid. Warm the mixture then pour it into about $15\,cm^3$ of water in a small beaker and smell the vapour.

5.35 Substance I

1 Heat a few crystals of substance I on a crucible lid, gently at first and then finally bring them to red heat.

2 Put a spatula load of substance I in a test-tube and then add $2\,cm^3$ of concentrated sulphuric acid; carefully heat the mixture and identify any gases evolved.

3 Put one spatula load of substance I in a flask and add about $20\,cm^3$ of sodium hydroxide solution and a few anti-bumping granules; boil under reflux for 15 minutes. Allow the apparatus to cool for five minutes before rearranging it for distillation: obtain $5\,cm^3$ of distillate.
a) Heat for a minute $2\,cm^3$ of the distillate with a spatula load of salicylic (2-hydroxybenzoic) acid and a few drops of concentrated sulphuric acid. Pour the mixture into about $15\,cm^3$ of water in a small beaker and smell the vapour.
b) Cool the residue in the flask and acidify it with dilute sulphuric acid. To $2\,cm^3$ of the mixture add potassium permanganate [manganate(VII)] solution dropwise until it is pink and then warm the coloured solution.

5.36 Substance J

Use one spatula load of substance J for each of the following tests.

1 Add $1\,cm^3$ of concentrated sulphuric acid and then carefully warm the mixture.

2 Add $4\,cm^3$ of sodium hydroxide solution and then carefully warm the mixture.

3 Add $4\,cm^3$ of dilute sulphuric acid, heat the mixture to boiling and then add potassium permanganate [manganate(VII)] solution dropwise.

4 Add $4\,cm^3$ of sodium hydroxide solution and then one drop *only* of copper(II) sulphate solution.

5 Heat some crystals of substance J on a crucible lid.

5.37 Substance K

1 Heat some crystals of substance K on a crucible lid.

2 Dissolve a few crystals of substance K in $2\,cm^3$ of water and then add $1\,cm^3$ of dilute nitric acid and $1\,cm^3$ of silver nitrate solution. Centrifuge and to the residue add $4\,cm^3$ of ammonia solution.

3 Dissolve half a spatula load of substance K in the minimum quantity of water, then add sodium hydroxide solution and finally an excess of dilute hydrochloric acid.

4 Dissolve some crystals of substance K in $3\,cm^3$ of dilute hydrochloric acid, add a few drops of sodium nitrite [nitrate(III)] solution and then gently boil the solution for five minutes.

5 Dissolve a few crystals of substance K in $2\,cm^3$ of water and then add sodium hypochlorite [chlorate(I)] solution dropwise.

VI Quantitative Analysis

Introduction

In qualitative analysis the identities of substances present are found. In quantitative analysis the proportions of substances present are found. Qualitative analysis of an unknown substance must precede quantitative analysis. Quantitative analysis can be divided into two parts:
1) gravimetric — weighings are done throughout
2) volumetric (or titrimetric) — volumes are measured mainly, possibly after one weighing.

Any reaction which takes place in solution can be followed if there are no side reactions and if the end-point, i.e. the completion of the reaction, is readily detectable.

The advantages of volumetric analysis compared with gravimetric analysis are that:
1) there are no adsorption or solubility losses
2) it is often accurate enough
3) the methods are more selective
4) it is quicker.

There are four types of estimation commonly performed by volumetric analysis:

1 **Acid-base** reactions (acidimetry, alkalimetry). This usually involves the neutralization of a base with an acid, both in aqueous solution; this is essentially
$$H^+ + OH^- \rightarrow H_2O$$
but the ionic equation may not be informative enough for many purposes. The end-point or equivalence point is made evident by the use of indicators: substances which show by their colour the nature of a solution. Neutralization can be detected also by measurements of the electrical conductance of a solution.

Strong acids are those which are highly ionized in solution (hydrochloric, nitric, sulphuric, etc.) whereas weak acids are only partially ionized [acetic (ethanoic), benzoic, oxalic (ethanedioic), etc.]. Strong bases are highly ionized in solution (sodium hydroxide, potassium hydroxide etc.) whereas weak bases are only partially ionized (ammonia, sodium carbonate, etc.). At the end-point of a reaction the solution is not always neutral, i.e. it is not always at pH = 7, because of hydrolysis of all salts except those formed by strong acids with strong bases. Hence the indicator must be chosen with care. The four cases are:

Strong acid + strong base	— any common indicator except litmus which is not sensitive enough for advanced work.
Strong acid + weak base	— methyl red, methyl orange and screened methyl orange.
Weak acid + strong base	— phenolphthalein.
Weak acid + weak base	— no suitable indicator.

These indicators are themselves weak acids. For a discussion of the change in pH during the course of a titration refer to a theory book.

2 **Redox** reactions (reduction and oxidation). The transfer of electrons is emphasized here and partial equations may be written for the oxidizing and reducing agent, e.g.
$$MnO_4^- + 8H^+ + 5e^- \rightarrow Mn^{2+} + 4H_2O$$
$$Fe^{2+} \rightarrow Fe^{3+} + e^-$$
Potassium permanganate functions as its own indicator but in most other cases an indicator must be added. Redox potential tables (an extension of the standard electrode potential tables encountered earlier) should be consulted here. The indicators used are often weak reducing agents.

3 **Precipitation** reactions (ionic association), e.g.
$$Ag^+ + X^- \rightarrow AgX\downarrow$$
$$(X = Cl, Br, I)$$
The solubility products of substances are important here. The indicators may be substances which also give precipitates or may function by adsorption.

4 **Complexometric** reactions, e.g.
$$M \quad + \quad nL \quad \rightleftharpoons ML_n$$
(usually an ion) (a ligand — may (complex)
be an ion)

The stability constant of the complex formed, i.e. the equilibrium constant of the appropriate reaction, is of importance here.

Standard solutions

A *standard solution* is one of which the concentration is accurately known.

A *1M solution* is one which contains the relative molecular (formula) mass of the substance in grams in a cubic decimetre (1 dm^3 or 1 litre = 1000 cm^3) of solution.

The relative molecular (atomic) mass in grams is called the molar mass; its units are g/mol.
Thus V cm^3 of a solution of concentration c mol/dm^3 contains $\frac{Vc}{1000}$ moles. The numbers of moles of two substances found to react with one another, if each is multiplied by the Avogadro constant, gives the numbers of molecules of the two substances reacting, which are in the same proportion as those in the equation. Thus if the equation is
$$aA + bB \rightarrow cC + dD$$
then
$$\frac{V_A c_A}{V_B c_B} = \frac{a}{b}$$
the other numbers involved cancelling out.

It is the equation for the reaction in the titration that must be used for this relationship and not the equation for any preliminary reactions.

Criteria for a primary standard

Highly pure substances are required in order to make standard solutions accurately. The following conditions should be fulfilled.

1) They must be easy to prepare, purify and preserve. Preferably they should withstand drying at 380K (107 °C)
2) They should not be hygroscopic, deliquescent or efflorescent.
3) They should be readily soluble in water.
4) The relative molecular mass should be reasonably high to minimize weighing errors but not so high that condition (3) is difficult to obtain. A suitable range of relative molecular masses is 50–500.
5) They should be capable of being tested for impurities.
6) Their reactions with other substances should be quantitative and practically instantaneous.

Primary and secondary standards are quoted in some books: ICI take silver as their ultimate standard, with iodine, sodium carbonate and sodium chloride as working standards and potassium dichromate as a secondary working standard. Other substances recommended as primary standards include hydrochloric acid (the constant boiling-point mixture), potassium chloride, potassium bromide, potassium iodate [iodate(V)], silver nitrate and sodium oxalate (ethanedioate); and as secondary standards, disodium tetraborate(III)-10-water, oxalic (ethanedioic) acid-2-water and copper(II) sulphate-5-water.

Volumetric apparatus

This is also a series of instructions for making up a standard solution.

The *weighing bottle* is a polystyrene container with a snap-on polythene lid. It can be used for inorganic substances and some organic ones but must be replaced by a glass one for many organic substances. It should be rinsed with distilled water and then dried with a paper towel; glass bottles are best dried in an oven. The approximate mass of the weighing bottle should be known (record it here: . . . . . g). The approximate quantity of the substance required should be loaded by spatula from the supply bottle into the weighing bottle (this should be done to an accuracy of ± 10%) and then the precise mass determined. As much as possible of the substance is then shaken into a clean (it may be wet still) 250 cm³ beaker and the bottle reweighed. All the weighings must be carefully recorded. Because the error in using a pipette is 1 in 400 there is no point in weighing out to a very much greater accuracy.

The substance is then dissolved in water in the beaker. If heating is necessary to hasten dissolution then the solution must be cooled before proceeding further. Do not put a stirring rod wet with solution down on the bench because this will cause loss of substance.

The *measuring (standard) flask* is made to contain a certain volume of solution at a particular temperature. The flask should be rinsed with distilled water and then the solution (at room temperature) poured in from the beaker. The beaker, stirring rod and funnel (if used) should be rinsed with distilled water and the rinsings added to the solution in the flask. The rinsing should be repeated at least once. The solution is then made up to the specified volume: the meniscus should just rest on the graduation mark thus:

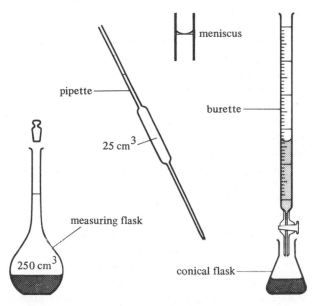

The solution should be shaken thoroughly to make it homogeneous.

The *pipette* is made to deliver, without blowing, a certain volume of solution at a particular temperature. The pipette should be rinsed with distilled water and then with the solution which it is about to contain. The solution is then sucked up to beyond the graduation mark: it is preferable to use a pipette filler and essential to use it if the solution is poisonous, corrosive or radioactive. The solution level is then allowed to run down to the graduation mark (see above). A conical flask is rinsed with water; it does not need to be dried. The tip of the pipette is allowed to touch the solution in the flask as the solution drains out of the pipette but despite this a small drop of solution always remains in the pipette. The graduation of a pipette allows for this remaining drop. Two drops, occasionally more, of an indicator are then added to the solution in the conical flask: the quantity of indicator employed in a particular set of titrations must be constant.

The *burette* is also made for delivering a volume of solution at a particular temperature. The burette should be rinsed with distilled water and then with the solution which it is about to contain — not forgetting the jet. The burette is then filled to above the zero mark; if a funnel is used it should be removed before proceeding further. The solution is then allowed to run down to the zero mark thus filling the jet. The meniscus should be as above: a piece of white paper held at an angle behind the burette may facilitate reading it.

The conical flask should be stood on a white tile for the titration. A rough titration should be performed first (proceeding 1 cm^3 at a time) and then several accurate titrations performed; two accurate titrations are essential and more may have to be done if the results are variable – an upward or downward trend is very suspect. The reasons for not obtaining concordant titrations include the following.

1) The solutions are not homogeneous.

2) The indicator is not used consistently either in amount or as regards extent of colour change.

3) There are atmospheric effects or gaseous products, e.g. carbon dioxide, which are not expelled from the solution.

4) The reaction is slow, e.g. potassium permanganate [manganate(VII)] reacts at a reasonable rate with an oxalate (ethanedioate) only above 340 K (67 °C).

By convention, if solution A is to be titrated with solution B, then solution B is put in the burette. All burette readings (start, finish and hence difference) must be recorded.

The apparatus should be rinsed after use:

> acids – by water
>
> alkalis – by dilute hydrochloric acid then water
>
> potassium permanganate – by acidified sodium sulphite solution   solution then water
>
> silver nitrate solution – by ammonia solution then water

Numerical error

The accuracy of the instruments used in volumetric analysis is

pipette	1 in 400
burette	1 in 500
measuring flask	1 in 800
weighing	1 in 1000

and to this must be added a personal error of at least 13 parts in 4000 making a total reasonable error of 1 in 100. Thus whilst four significant figures should be carried in calculations (no greater approximations allowed) the final answer should be quoted to three significant figures. If, however, the sequence of digits is above 500 or it is realized that a particular experiment cannot be accomplished with high accuracy then a more realistic answer is to two significant figures.

6.1 The Standardization of Hydrochloric Acid with Sodium Carbonate

1 Na_2CO_3 + $2HCl$ → $2NaCl + H_2O + CO_2\uparrow$
   106.0 g (anhydrous) $2\,dm^3$ 1M solution
   Mass for $250\,cm^3$ of 0.05 M sodium carbonate solution ≈ 1.3 g.
   Weigh this out and make up $250\,cm^3$ solution according to the general procedure. Record all masses carefully.

2 A suitable sample of acid for studying is that supplied to the laboratory (probably 35% m/V) diluted 100 times.

3 Titrate $25\,cm^3$ portions of the sodium carbonate solution with the diluted acid using a suitable indicator (for this purpose sodium carbonate behaves as a weak base). Slight warming of the alkali facilitates the expulsion of carbon dioxide; alternatively 0.2M acid may be used.

4 Calculate the concentration of the diluted acid and hence that of the original acid.

6.2 The Standardization of Hydrochloric Acid with Disodium Tetraborate(III)–10–Water

1 $Na_2B_4O_7$ + $2HCl + 5H_2O$ → $2NaCl + 4H_3BO_3$
   381.4 g (as $10H_2O$) $2\,dm^3$ 1M solution   Boric acid
   Mass for $250\,cm^3$ 0.05M disodium tetraborate (borax) solution ≈ 4.8 g.
   Weigh this out and make up $250\,cm^3$ solution according to the general procedure. Record all masses carefully. Warming may be necessary to hasten dissolution of the crystals but afterwards the solution must be carefully cooled back to room temperature. Boric acid does not affect methyl orange.

2 A suitable sample of acid for studying is that supplied to the laboratory (probably 35% m/V) diluted 100 times.

3 Titrate $25\,cm^3$ portions of the borax solution with the diluted acid using screened methyl orange as the indicator.

4 Calculate the concentration of the diluted acid and hence that of the original acid.

6.3 The Determination of the Purity of an Insoluble Base

1 $CaCO_3$ + $2HCl$ → $CaCl_2 + H_2O + CO_2\uparrow$
   100.1 g (e.g. chalk) $2\,dm^3$ 1M solution
   Mass of calcium carbonate to react with $25\,cm^3$ 1M hydrochloric acid ≈ 1.3 g.
   Weigh this out and put it in a conical flask.

2 By pipette put $50\,cm^3$ of 1M hydrochloric acid on to the calcium carbonate. The reason for this is that on dilution to $250\,cm^3$ the excess acid will be approximately 0.1M. When the evolution of carbon dioxide is completed pour the contents of the conical flask into a measuring flask and add the rinsings. Make up the volume to $250\,cm^3$.

3 Titrate $25\,cm^3$ portions of the acidic solution with standard alkali solution, e.g. 0.1M sodium hydroxide solution, using a suitable indicator.

4 The calculation can be done in terms of millimoles of reagents. Calculate the concentration (cM) of the excess acid from the titration results, and hence the number of millimoles of acid left after the reaction is $250c$. The number of millimoles of acid taken at the start is 50×1. Hence the number of millimoles of acid that have reacted is $50-250c$. From the equation two millimoles of acid are equivalent to one millimole, i.e. 0.1001 g, of pure calcium carbonate. Thus the mass of pure carbonate taken can be calculated and then the percentage purity.
   This type of estimation is known as back titration.

6.4 The Determination of the Proportion of Nitrogen in a Fertilizer

1

$$NH_4^+ + OH^- \rightarrow H_2O + NH_3 \uparrow$$

e.g. $(NH_4)_2SO_4 + 2NaOH \rightarrow Na_2SO_4 + 2H_2O + 2NH_3 \uparrow$
         132.1 g   2 dm³ 1M
                      solution

Mass of ammonium sulphate to react with 25 cm³ 1M sodium hydroxide solution $\approx$ 1.7 g.
Weigh this out and put it in a conical flask.

2 By pipette put 50 cm³ of 1M sodium hydroxide solution on to the ammonium sulphate; see experiment 6.3 for the reasoning. Boil the mixture until no more ammonia is evolved: do not let the flask run dry. A glass funnel in the neck of the flask may be advantageous but it must be rinsed inside and outside afterwards. Cool the flask and contents back to room temperature and pour the solution, together with rinsing water, into a measuring flask and make up the volume to 250 cm³. A fume cupboard may be used when the ammonia is boiled off.

3 Titrate 25 cm³ portions of the alkaline solution with a standard acid, e.g. 0.1M hydrochloric acid, using a suitable indicator.

4 The calculation is similar to that in experiment 6.3. Here 1 mole sodium hydroxide $\equiv$ 14.01 g nitrogen Hence calculate the percentage of nitrogen in the fertilizer. Compare the value obtained with that calculated from the appropriate relative atomic and molecular masses.
    This exercise is another example of back titration.

6.5 A Double Indicator Experiment

e.g. The analysis of a mixture of sodium hydroxide and sodium carbonate.

1 a) $NaOH + HCl \rightarrow NaCl + H_2O$
   b) $Na_2CO_3 + HCl \rightarrow NaHCO_3 + NaCl$
   c) $NaHCO_3 + HCl \rightarrow NaCl + H_2O + CO_2 \uparrow$
   The reaction of sodium hydroxide solution with hydrochloric acid is followed using phenolphthalein as the indicator. The reaction of sodium carbonate solution proceeds in two stages, the first being marked by phenolphthalein just changing from pink to colourless and the second by screened methyl orange changing colour. A standard acid, e.g. 0.2M hydrochloric acid, is used for the titration.

2 Titrate 25 cm³ portions of the alkaline solution with the acid using phenolphthalein as the indicator. When the colour change has occurred, read the burette, add two drops of screened methyl orange to the mixture in the conical flask and continue the titration, reading the burette a third time when the second indicator changes colour. Let the first titre be T_1 and the total titre be T_2 cm³ (the titre is the volume of solution added from the burette).

3 The first titre (T_1 cm³) measures the quantity of acid for reactions (a) and (b); the difference ($T_2 - T_1$ cm³) measures the quantity of acid for reaction (c). The sodium hydrogencarbonate is a common factor in reactions (b) and (c) so the quantity of acid to react with the sodium carbonate solution completely is $2(T_2 - T_1)$ cm³. Thus the quantity of acid to react with the sodium hydroxide solution is $T_2 - 2(T_2 - T_1)$, i.e. $2T_1 - T_2$ cm³.

4 The alkalis may be considered to be existing separately in the 25 cm³ solution taken and so their concentrations (in g/dm³) may now be calculated.

6.6 Winkler's Experiment

1 The solutions for analysis may be the same as in experiment 6.5 so that the two approaches to the problem may be compared.

2 The total titration of the alkalis in solution is done using screened methyl orange as the indicator; if the two experiments are being compared the total titre T_2 cm³ may be assumed here.

3 Take a new 25 cm³ portion of the alkaline solution and add an excess of barium chloride solution (about 25 cm³ 0.1M solution) to precipitate all carbonate ions.
   $$Ba^{2+} + CO_3^{2-} \rightarrow BaCO_3 \downarrow$$
   This leaves sodium hydroxide only in solution and if phenolphthalein is added as the indicator a titration with standard hydrochloric acid can be performed with much swirling of the flask to prevent any local excesses of acid dissolving any precipitate. Let the titre be T_3 cm³.

4 This titre (T_3 cm³) when subtracted from the total titre (T_2 cm³) gives the quantity of acid which reacts with the sodium carbonate solution.

5 The alkalis may be considered to be existing separately in the 25 cm³ solution taken and so their concentrations (in g/dm³) may now be calculated.

6.7 The Determination of Water of Crystallization in a Hydrate

1 $Na_2CO_3 + 2HCl \rightarrow 2NaCl + H_2O + CO_2 \uparrow$
Heating the crystals to constant mass shows that there is a large proportion of water of crystallization. If the formula is taken to be $Na_2CO_3 \cdot 7H_2O$ then the mass for $250\ cm^3$ 0.05 M solution ≈ 3 g.
   Weigh this out and make up to $250\ cm^3$ solution.

2 Titrate $25\ cm^3$ portions of the sodium carbonate solution with a standard acid, e.g. 0.1M hydrochloric acid, using a suitable indicator. (An alternative to warming the solution is to use 0.2M hydrochloric acid).

3 Calculate the concentration of the alkali in mol/dm^3. Then if the crystals have the formula $Na_2CO_3 \cdot xH_2O$

$$M \text{ concentration} = \frac{\text{Mass of crystals in } 1\ dm^3 \text{ solution}}{\text{Molar mass (g/mol)}}$$

$$= \frac{4 \times \text{mass taken in grams}}{106.0 + 18.01x}.$$

Calculate x.

6.8 The Analysis of a Soda Mint Tablet

1 The main constituent of a soda mint tablet is sodium hydrogencarbonate; this method ascribes all the alkalinity of the tablet to be due to this salt.

2 $NaHCO_3 + HCl \rightarrow NaCl + H_2O + CO_2 \uparrow$
   84.01 g $\quad$ $1\ dm^3$ 1M
   $\qquad\qquad$ solution
Weigh out one tablet and put it into a conical flask containing about $25\ cm^3$ of distilled water, crush the tablet using a glass rod; add two drops of screened methyl orange.

3 Titrate the alkaline solution with standard (e.g. 0.1M) hydrochloric acid.

4 Repeat the steps (2) and (3) with another tablet. Calculate the average value of $\frac{\text{titre}}{\text{mass of tablet}}$. Hence calculate the mass of pure sodium hydrogencarbonate in the tablet and the percentage by mass.

6.9 The Analysis of Lemon Squash

1 When citric acid HO$-$C$-$COOH $\cdot H_2O$ with CH$_2$COOH groups
(2-hydroxypropane-1,2,3-tricarboxylic acid) is titrated, using phenolphthalein as the indicator, it behaves as a dibasic acid. Write the equation for the reaction.

2 Titrate $10\ cm^3$ of lemon squash and about $15\ cm^3$ of distilled water with a standard (e.g. 0.1M) solution of sodium hydroxide in the presence of two drops of phenolphthalein. Rinse the burette thoroughly when you have finished.

3 Calculate the percentage mass/volume of acid in the squash.

6.10 The Determination of the Relative Molecular Mass of a Dibasic Organic Acid

1 What is the range of the relative molecular masses of common simple substances?

2 For your titration you might need to prepare a solution which is roughly 0.05M (0.5×0.1 because it is dibasic). Is the substance supplied soluble enough to make a solution of this concentration? If so, a direct titration can be performed; if not, a back titration must be done.

3 Should the acid solution be titrated with a standard (e.g. 0.1M) solution of sodium hydroxide or ammonia; or can either be used? The acid is organic: so it may be either a strong or a weak acid. Which indicator must be used?

4 If a simple titration is to be done titrate $25\ cm^3$ of the standard alkali supplied with the acid solution. If a back titration is done 0.0125 moles (0.025×0.5 because it is dibasic) of material are weighed out and dissolved in $50\ cm^3$ of standard (e.g. 1M) sodium hydroxide solution and the mixture diluted to $250\ cm^3$ before $25\ cm^3$ portions are titrated with standard (e.g. 0.1M) acid.

6.11 The Analysis of a Lavatory Cleanser

1 Qualitative analysis shows that 'Harpic', 'Sanilav' etc. contain sodium hydrogensulphate and this experiment ascribes all the acidity as being due to this salt.

2 $NaHSO_4 + NaOH \rightarrow Na_2SO_4 + H_2O$
   120.1 g   1 dm³ 1M
              solution

   A solution containing 12 g/dm³ of a lavatory cleanser is supplied. Titrate 25 cm³ portions of a standard (e.g. 0.1M) sodium hydroxide solution with the unknown solution using phenolphthalein as the indicator.

3 Calculate the percentage by mass of sodium hydrogensulphate in the lavatory cleanser supplied.

6.12 The Analysis of Dolomite

Dolomite is calcium magnesium carbonate.

1 You are given (a) hydrochloric acid (approximately 0.1M)
   (b) sodium carbonate solution (0.075M)
   (c) hydrochloric acid which contains 3.7 g/dm³ of dolomite (a different sample of hydrochloric acid)

2 Titrate 25 cm³ of (a) with (b) using screened methyl orange as the indicator. Calculate the exact concentration of (a).

3 Titrate 25 cm³ of (c) with (b) using screened methyl orange as the indicator. Calculate the concentration of the residual acid in (c).

4 To 25 cm³ of (c) in a conical flask add 50 cm³ of (b). Heat the mixture carefully until it is nearly boiling and then allow it to cool. Filter off the precipitate while the solution is still warm and rinse the precipitate several times with distilled water: add the rinsings to the filtrate. Titrate the filtrate and rinsings with (a) using screened methyl orange as the indicator.

5 Calculate the masses of magnesium carbonate and calcium carbonate dissolved in the acid in (c). Hence calculate the formula of dolomite.

6 What is the concentration of the hydrochloric acid originally supplied for solution (c)?

6.13 Acidimetry Problems

1 **The determination of the solubility of calcium hydroxide**
   The solubility of calcium hydroxide in water may be found by titration with 0.05 M hydrochloric acid. If hot solutions are filtered at various temperatures then the solubility curve can be drawn.

2 **An alternative method of studying an ammonium salt**
   Boil the ammonium salt, e.g. 1.7 g ammonium sulphate, with 50 cm³ of 1M sodium hydroxide solution and pass the vapours into 50 cm³ 0.5M sulphuric acid via an inverted funnel. Make up the residual sulphuric acid to 250 cm³ and titrate 25 cm³ portions with standard (e.g. 0.1M) sodium hydroxide solution using screened methyl orange as the indicator.

3 **The determination of the relative atomic mass of magnesium**
   Dissolve 0.3 g of clean magnesium ribbon in 50 cm³ of 1M hydrochloric acid, dilute the residual acid to 250 cm³ and titrate 25 cm³ portions with alkali as in (2).

4 **The determination of the purity of calcium oxide**
   As (3) using 0.7 g calcium oxide.

5 **The determination of the water content of glacial acetic acid**
   Dilute 1 cm³ of glacial acetic acid (pure ethanoic acid, density 1.05 g/cm³) to 250 cm³ and titrate 25 cm³ portions with standard (e.g. 0.1M) sodium hydroxide solution.

6 **The determination of the water of crystallization of oxalic acid**
   Oxalic (ethanedioic) acid crystals have the formula $H_2C_2O_4 \cdot xH_2O$ and the acid is dibasic. You are given a standard alkaline solution. Proceed with due caution because oxalates are poisonous.

Redox Titrations: (a) Permanganate

Introduction

In reactions an oxidizing agent gains electrons, these being supplied by the reducing agent. The processes of reduction and oxidation are complementary, i.e. one cannot occur without the other, hence the term redox is a reasonable abbreviation. The molecular equation for a redox reaction can be studied first, to yield the ionic equation by cancelling out ions which appear both amongst reactants and products; secondly, partial equations can be constructed for the oxidizing and reducing agents.

Consider the reaction

$$2KMnO_4 + 8H_2SO_4 + 10FeSO_4$$
$$\rightarrow K_2SO_4 + 2MnSO_4 + 5Fe_2(SO_4)_3 + 8H_2O$$

converting into ions, where relevant,

$$2K^+ + 2MnO_4^- + 16H^+ + 8SO_4^{2-} + 10Fe^{2+} + 10SO_4^{2-}$$
$$\rightarrow 2K^+ + SO_4^{2-} + 2Mn^{2+} + 2SO_4^{2-} + 10Fe^{3+} + 15SO_4^{2-} + 8H_2O$$

cancelling out ions where possible and dividing by two,

$$2MnO_4^- + 16H^+ + 10Fe^{2+} \rightarrow 2Mn^{2+} + 10Fe^{3+} + 8H_2O$$
$$MnO_4^- + 8H^+ + 5Fe^{2+} \rightarrow Mn^{2+} + 5Fe^{3+} + 4H_2O.$$

On constructing partial equations:

$$MnO_4^- + 8H^+ + 5e^- \rightarrow Mn^{2+} + 4H_2O$$

Purple → Colourless in solution

$$Fe^{2+} \rightarrow Fe^{3+} + e^-$$

In practice on future occasions, it is better to work in the opposite direction constructing first the partial equations and then obtaining the ionic and/or molecular equations if required.

Potassium permanganate [manganate(VII)] is usually employed under acidic conditions. Potassium permanganate is an oxidizing agent towards hydrochloric acid (especially the concentrated acid) whilst nitric acid is also an oxidizing agent, so the acid most suitable for creating an acidic solution is sulphuric. A 0.02 M solution of potassium permanganate may be used: the permanganate equation involves five electrons and so the concentration for comparability is one fifth of 0.1M. The permanganate is put in the burette and added to 25 cm³ of the reducing agent and 20 cm³ of 1M sulphuric acid in the conical flask until the solution assumes a permanent pale pink tinge. If a brown precipitate of manganese(IV) oxide appears acidification is insufficient and that titration may have to be repeated.

For a discussion of the changes in redox potential during the course of a titration refer to a theory book. In most redox titrations an indicator has to be added. Simple iron(II) salts are too unstable to be used as standardizing agents and usually ammonium iron(II) sulphate, a double salt, or an oxalate (ethanedioate) is employed.

Occasionally potassium permanganate is used under neutral or alkaline conditions, the partial equation being

$$MnO_4^- + 2H_2O + 3e^- \rightarrow MnO_2\downarrow + 4OH^-$$

Purple → Brown probably hydrated

The advantages of potassium permanganate as an oxidizing agent are as follows.
1) It acts as its own indicator because the manganese(II) ions are too pale a pink to impart a noticeable colour to the solution.
2) The crystals are obtainable in a high state of purity.
3) The crystals are anhydrous and not deliquescent.
4) The compound has a fairly high relative molecular mass.
5) With it, a wide range of substances may be oxidized quantitatively.

On the other hand, potassium permanganate does have some disadvantages, as shown.
1) The crystals are not very soluble in water.
2) The compound is reduced by organic matter from the atmosphere.
3) The compound is decomposed by light.
4) The meniscus of the solution may be difficult to see.

6.14 The Standardization of Potassium Permanganate with an Iron(II) Salt

1 $MnO_4^- + 8H^+ + 5Fe^{2+} \rightarrow Mn^{2+} + 5Fe^{3+} + 4H_2O$

The relative molecular mass of ammonium iron(II) sulphate-6-water is 392.1 so the mass to be taken for $250\,cm^3$ of 0.1M solution $\approx 10\,g$.

The crystals should be made into a standard solution without any heating; about $50\,cm^3$ 1M sulphuric acid should be included to reduce hydrolysis and oxidation. In this reaction the ammonium sulphate of the double salt is not affected at all.

2 Titrate $25\,cm^3$ portions of the iron(II) solution, to which about $20\,cm^3$ of 1M sulphuric acid are added each time, with potassium permanganate solution until a permanent pale pink tinge is seen.

3 Calculate the concentration of the potassium permanganate solution.

6.15 The Standardization of Potassium Permanganate with an Oxalate (Ethanedioate)

1 $2MnO_4^- + 16H^+ + 5C_2O_4^{2-} \rightarrow 2Mn^{2+} + 8H_2O + 10CO_2\uparrow$

The relative molecular mass of sodium oxalate is 134.0 so the mass to be taken for $250\,cm^3$ of 0.05M solution $\approx 1.6\,g$.

Oxalic acid is an alternative to the salt; it is usually used in the form of its dihydrate. **Care:** oxalates and the acid are poisonous — a pipette filler is thus essential for this experiment.

2 The reaction does not proceed at a reasonable rate below 340K (67°C); warm a $25\,cm^3$ portion of the oxalate solution with about $20\,cm^3$ of 1M sulphuric acid until the flask is just too hot to hold comfortably. Titrate the solution as rapidly as possible with potassium permanganate solution, rewarming if necessary.

3 Calculate the concentration of the potassium permanganate solution.

6.16 The Determination of the Water of Crystallization of an Iron(II) Salt

1 $MnO_4^- + 8H^+ + 5Fe^{2+} \rightarrow Mn^{2+} + 5Fe^{3+} + 4H_2O$

The relative molecular mass of hydrated iron(II) sulphate is $151.9 + 18.01\,x$. By taking $x = 6$ as being a possible answer, the mass of crystals to be taken for $250\,cm^3$ of 0.1M solution is $\approx 6.4\,g$.

The crystals should be made into a standard solution without any heating; about $50\,cm^3$ of 1M sulphuric acid should be included to reduce hydrolysis and oxidation.

2 Titrate $25\,cm^3$ portions of the iron(II) solution, to which about $20\,cm^3$ of 1M sulphuric acid are added each time, with standard (e.g. 0.02M) potassium permanganate solution until a permanent pale pink tinge is seen.

3 Calculate the concentration of the iron(II) solution and hence x, the number of molecules of water of a crystallization per molecule of iron(II) sulphate.

$$M \text{ concentration} = \frac{\text{Mass of crystals in } 1\,dm^3 \text{ solution}}{\text{Molar mass (g/mol)}}$$

$$= \frac{4 \times \text{mass taken in grams}}{151.9 + 18.01\,x}$$

6.17 The Determination of the Purity of an Iron Wire

1 $Fe + 2H^+ \rightarrow Fe^{2+} + H_2\uparrow$

$MnO_4^- + 8H^+ + 5Fe^{2+} \rightarrow Mn^{2+} + 5Fe^{3+} + 4H_2O$

The relative atomic mass of iron is 55.85 so the mass of iron (or steel) wire for $250\,cm^3$ 0.1M solution $\approx 1.4\,g$. The wire should be weighed out and added to about $75\,cm^3$ of 1M sulphuric acid in a conical flask: a funnel may be put into the mouth of the conical flask to avoid loss by spurting. Gently heat the flask and leave it in a fume cupboard until the reaction is complete (some fine black particles of carbon may be left in suspension). The funnel should be rinsed inside and outside and the rinsings added to the solution. The solution should be cooled back to room temperature and then poured into a $250\,cm^3$ measuring flask, the rinsings from the conical flask being added, and the solution made up to the mark.

2 Titrate $25\,cm^3$ portions of the iron(II) solution, to which about $20\,cm^3$ of 1M sulphuric acid are added each time, with standard (e.g. 0.02M) potassium permanganate solution until a permanent pale pink tinge is observed.

3 Calculate the concentration in mol/dm^3 of the iron(II) solution and multiply this by the relative atomic mass of iron (the mass of two electrons being negligible). Hence the percentage purity of the iron wire can be found.

6.18 The Analysis of Hydrogen Peroxide Solution

1 $2MnO_4^- + 6H^+ + 5H_2O_2 \rightarrow 2Mn^{2+} + 8H_2O + 5O_2 \uparrow$
   A suitable sample of hydrogen peroxide solution for study is that supplied to the laboratory (50-volume) diluted 100 times.

2 Titrate 25 cm^3 portions of the peroxide solution, to which about 20 cm^3 of 1M sulphuric acid are added each time, with standard (e.g. 0.02M) potassium permanganate solution until a permanent pale pink tinge is seen.

3 Calculate the concentration in g/dm^3 and mol/dm^3 of the original hydrogen peroxide solution.

4 The volume strength of hydrogen peroxide is calculated according to the equation for its thermal or catalytic decomposition.
   $$2H_2O_2 \rightarrow 2H_2O + O_2 \uparrow$$
   68.02 g   22.41 dm^3
   2 moles

   If this 68.02 g of hydrogen peroxide is in 1 dm^3 solution then, because it yields 22.41 dm^3 oxygen, it is called a 22.41-volume solution. A 1M solution is an 11.2 volume solution. Hence calculate the volume strength of the hydrogen peroxide supplied to the laboratory.

6.19 The Determination of the Proportion of Iron(III) in a Salt

1           $2Fe^{3+} + Zn \rightarrow 2Fe^{2+} + Zn^{2+}$
   $MnO_4^- + 8H^+ + 5Fe^{2+} \rightarrow Mn^{2+} + 5Fe^{3+} + 4H_2O$
   A suitable substance to study is ammonium iron(III) sulphate-12-water which has a relative molecular mass of 482.2, so the mass to be taken for 250 cm^3 of a 0.1M solution $\approx$ 12 g. The crystals should be made up into a solution which includes about 75 cm^3 of 1M sulphuric acid to reduce hydrolysis.

2 To four separate 25 cm^3 portions of the iron(III) solution in conical flasks add about 25 cm^3 of 1M sulphuric acid, about 1 cm^3 of 1M copper(II) sulphate solution and about 1–2 g of zinc dust (or the granulated metal). When a drop of the solution withdrawn on a glass rod no longer gives a deep red coloration with potassium thiocyanate solution the reduction is nearly completed and a rough titration can be performed. For more accurate titrations it is necessary to wait for no coloration or only a very pale pink to be produced.
   $Fe^{3+} + SCN^- \rightarrow [FeSCN]^{2+}$

3 Filter each portion of the iron solution and rinse the remaining zinc, adding the rinsings to the filtrate. Titrate each batch with standard (e.g. 0.02M) potassium permanganate solution until a permanent pale pink tinge is seen.

4 Calculate the concentration in mol/dm^3 of the iron solution and multiply this by the relative atomic mass of iron (the mass of the electrons being negligible). Hence the percentage of iron in the salt chosen can be found.

6.20 The Study of an Acid Oxalate (Ethanedioate)

1 The acid oxalate supplied contains 8 g/dm^3 of $K_aH_b(C_2O_4)_c \cdot dH_2O$. Write equations involving b and c for the titrations of this salt with sodium hydroxide solution and potassium permanganate solution respectively.

2 By pipette and filler (**Care**: oxalates poisonous) put 25 cm^3 of the solution into a conical flask and titrate it with standard (e.g. 0.1M) sodium hydroxide solution using phenolphthalein as the indicator. Calculate the concentration of the oxalate solution as a function of b.

3 Take a new portion of the oxalate solution, add about 75 cm^3 of 1M sulphuric acid and heat the flask and contents until the flask is just too hot to hold comfortably. Titrate the solution as rapidly as possible, rewarming if necessary, with standard (e.g. 0.02M) potassium permanganate solution. Calculate the concentration of the oxalate solution as a function of c.

4 The oxalate solution used for the titrations is the same throughout so the concentrations calculated in sections (2) and (3) are equal: this enables you to calculate the ratio $b:c$ in the salt. By summing up the charges on the ions it is now possible to suggest a value for a.

5 Using the values of b and c obtained in section (4) it is possible to calculate two values for the concentration of the potassium hydrogenoxalate and these values may then be averaged. Finally
   $$M \text{ concentration} = \frac{\text{Mass of crystals in 1 dm}^3 \text{ solution}}{\text{Molar mass (g/mol)}}$$
   Hence d may be calculated.

6.21 A Problem Involving Potassium Permanganate

1 Titrate 25 cm³ portions of the iron(II) solution supplied, to which about 20 cm³ of 1M sulphuric acid are added each time, with the standard (e.g. 0.02M) potassium permanganate solution. Calculate the concentration of the iron(II) solution.

2 Titrate a 25 cm³ portion of the iron(II) solution, to which about 20 cm³ of 1M sulphuric acid and exactly 10 cm³ of X are added, with the permanganate solution.

3 Repeat (2) using successively 20, 30 and 40 cm³ of X in the flask.

4 Plot a graph of your titres in experiments (1)–(3) (vertical axis) against the volume of X used (horizontal axis) and comment upon its shape.

5 What volume of X reacts with 25 cm³ of the iron(II) solution?

6 What can you deduce about the nature of solution X?

7 Analyze solution X to confirm any suspicions you have about its identity.

   Write equations for all reactions involved and if possible calculate the concentration of X.

6.22 The Determination of the Purity of a Nitrite [Nitrate(III)]

1 $2MnO_4^- + 6H^+ + 5NO_2^- \rightarrow 2Mn^{2+} + 5NO_3^- + 3H_2O$
   The relative molecular mass of sodium nitrite is 69.00 so the mass to be taken for 250 cm³ of a 0.05M solution $\approx$ 0.8 g. This titration may be a preliminary exercise before a diazo salt is prepared. The nitrite solution should be put in the burette because it becomes too unstable upon acidification to be titrated in the usual manner for permanganate titrations.

2 Titrate 25 cm³ portions of a standard (e.g. 0.02M) potassium permanganate solution, to which about 20 cm³ of 1M sulphuric acid are added each time, with the nitrite solution, swirling the flask vigorously until the pink colour has just been discharged.

3 Calculate the concentration and hence the percentage purity of the sodium nitrite.

6.23 Permanganate Problems

1 **The determination of the purity of calcite**
   Dissolve 1.2 g calcite (a form of calcium carbonate) in dilute hydrochloric acid. Add ammonia until the solution is alkaline and then ammonium oxalate (ethanedioate) to precipitate calcium oxalate. Dissolve the calcium oxalate in the minimum quantity of dilute hydrochloric acid and make up the volume to 250 cm³ in a measuring flask. Titrate 25 cm³ portions of the solution, to which about 20 cm³ of 1M sulphuric acid are added each time, with standard (e.g. 0.02M) potassium permanganate solution at 340 K (67 °C, see experiment 6.15).
   Calculate the percentage purity of the calcite.

2 **The determination of the purity of pyrolusite**
   Dilute 100 cm³ oxalic (ethanedioic) acid solution (16 g/dm³) to 250 cm³ and titrate 25 cm³ portions, to which about 20 cm³ of 1M sulphuric acid are added each time, with standard (e.g. 0.02M) potassium permanganate solution at 340 K (67 °C, see experiment 6.15).
   To a second 100 cm³ portion of oxalic acid add about 20 cm³ of 1M sulphuric acid and 1 g pyrolusite (a form of manganese(IV) oxide); boil the mixture for a few minutes, cool it back to room temperature and make up the volume to 250 cm³ in a measuring flask. Titrate as above.
   Calculate the percentage of manganese(IV) oxide in pyrolusite.

Care: a pipette filler must be used in both of these problems.

Redox Titrations: (b) Dichromate

Introduction

The partial equation for a dichromate [dichromate(VI)] behaving as an oxidizing agent in acidic solution is

$$Cr_2O_7^{2-} + 14H^+ + 6e^- \rightarrow 2Cr^{3+} + 7H_2O$$
Orange   Green

A $0.01\dot{6}$ M solution of potassium dichromate may be used: the dichromate equation involves six electrons and so the concentration for comparison is one sixth of 0.1 M.

An indicator is vital in view of the colour change involved: the indicator is a weak reducing agent such as diphenylamine, or N-phenylanthranilic (N-phenyl-2-aminobenzoic) acid, both of which give a red-violet colour when there is excess dichromate present. To the mixture for titration (25 cm³ of the reducing agent and about 20 cm³ of 1M sulphuric acid) it is necessary to add 10 cm³ of 1M phosphoric acid to prevent iron(III) ions oxidizing the indicator – they are converted to $[FeHPO_4]^+$

When dichromate is compared to permanganate the following points should be noted.
1) It is a less vigorous oxidizing agent.
2) It is easier to obtain pure and can be fused safely to dry it.
3) It can be used in the presence of a higher concentration of chloride ions.
4) It does not react with oxalates (ethanedioates).

6.24 The Reaction of Potassium Dichromate with an Iron(II) Salt

1   $Cr_2O_7^{2-} + 14H^+ + 6Fe^{2+} \rightarrow 2Cr^{3+} + 6Fe^{3+} + 7H_2O$
Read the specifications for the potassium dichromate [dichromate(VI)] and ammonium iron(II) sulphate supplied to the laboratory. It may be found that the dichromate is of higher purity than the iron(II) salt. The relative molecular mass of potassium dichromate is 294.2 so the mass to be taken for 250 cm³ of $0.01\dot{6}$ M solution ≈ 1.2 g. The relative molecular mass of ammonium iron(II) sulphate-6-water is 392.1 so the mass to be taken for 250 cm³ of 0.1M solution ≈ 10 g (see experiment 6.14).

2   Titrate 25 cm³ portions of the iron(II) salt solution, to which about 20 cm³ of 1M sulphuric acid, about 10 cm³ of 1M phosphoric acid and a few drops of diphenylamine are added each time, with the dichromate solution.

3   Compare the concentration of the iron(II) salt calculated from the weighings with that obtained from the titrations. Do the results agree with the manufacturer's specification?

6.25 The Determination of the Purity of Precipitated Copper

1   $3Cu + Cr_2O_7^{2-} + 14H^+ \rightarrow 3Cu^{2+} + 2Cr^{3+} + 7H_2O$
The relative atomic mass of copper is 63.54 so the mass to be taken for this experiment (to react with about 25 cm³ $0.01\dot{6}$M potassium dichromate solution) ≈ 0.1 g.

2   Warm the weighed portion of copper powder with 50 cm³ standard (e.g. $0.01\dot{6}$M) potassium dichromate solution and about 20 cm³ 1M sulphuric acid in a conical flask for about 20 minutes. Cool the mixture back to room temperature, add a few drops of indicator, 10 cm³ of 1M phosphoric acid and titrate with a standard iron(II) salt [e.g. 0.1M ammonium iron(II) sulphate] solution.

3   Calculate the percentage purity of the precipitated copper.

6.26 The Determination of the Purity of Tin

1
$$Sn + 2H^+ \rightarrow Sn^{2+} + H_2 \uparrow$$
$$Cr_2O_7^{2-} + 14H^+ + 3Sn^{2+} \rightarrow 2Cr^{3+} + 3Sn^{4+} + 7H_2O$$
The relative atomic mass of tin is 118.7 so the mass to be taken for this experiment (to react with about 250 cm³ 0.016M dichromate solution) ≈ 1.5 g.

The tin is dissolved in about 50 cm³ concentrated hydrochloric acid and the solution made up to 250 cm³ in a measuring flask.

2 Titrate 25 cm³ portions of the tin(II) solution, to which about 25 cm³ 0.2M iron(III) chloride solution (made up in 1M hydrochloric acid), about 10 cm³ of 1M phosphoric acid and a few drops of indicator are added each time, with standard (e.g. 0.016M) potassium dichromate solution. The initial reaction is
$$2Fe^{3+} + Sn^{2+} \rightarrow 2Fe^{2+} + Sn^{4+}$$

3 Calculate the concentration of the tin(II) solution and hence the purity of the tin.

6.27 A Study of the Reaction of Copper with an Iron(III) Salt

1 A possible equation for the reaction of copper with an iron(III) salt is
$$Cu + 2Fe^{3+} \rightarrow Cu^{2+} + 2Fe^{2+}$$
The problem is to find out whether the quantities of reactants used follow from this equation.

2 Warm 0.15 g copper powder (the purity of which has been determined in experiment 6.25) with about 50 cm³ of an iron(III) solution, e.g. 0.1M ammonium iron(III) sulphate made up in 1M sulphuric acid, in a conical flask for about 20 minutes. A Bunsen valve may be fitted to the flask to minimize possible oxidation of the products.

3 Cool the solution, add about 10 cm³ of 1M phosphoric acid and a few drops of indicator and titrate it with standard (e.g. 0.016M) potassium dichromate solution.

4 Does the equation quoted above describe the reaction quantitatively?

6.28 The Determination of the Concentration of Ethanol in Blood

A version of the 'alcohol in blood' test.

1
$$Cr_2O_7^{2-} + 8H^+ + 3C_2H_5OH \rightarrow 2Cr^{3+} + 3CH_3CHO + 7H_2O$$
$$Cr_2O_7^{2-} + 14H^+ + 6Fe^{2+} \rightarrow 2Cr^{3+} + 6Fe^{3+} + 7H_2O$$
To 25 cm³ of the 'blood' supplied add 25 cm³ of standard (e.g. 0.016M) potassium dichromate and then about 10 cm³ of concentrated sulphuric acid, with swirling of the flask. Heat the mixture strongly for five minutes and then cool it back to room temperature.

2 The cold solution contains an excess of potassium dichromate: to it add 25 cm³ of a standard iron(II) solution (e.g. 0.1M ammonium iron(II) sulphate), about 15 cm³ of 1M sulphuric acid, about 10 cm³ of 1M phosphoric acid and a few drops of diphenylamine. Titrate the excess of iron(II) salt with the standard dichromate solution; the amount required is the quantity of the dichromate which reacted with ethanol in stage (1).

Stages (1) and (2) should be done twice.

3 Calculate the concentration of ethanol in the blood. If the concentration is above 80 mg/100 cm³ (ml) the person cannot legally drive a vehicle in the United Kingdom.

Redox Titrations: (c) Iodine and Thiosulphate

Introduction

$$2S_2O_3^{2-} + I_2 \rightarrow S_4O_6^{2-} + 2I^-$$

Thiosulphate colourless — Iodine brown — Tetrathionate, Iodide — Colourless

$$2S_2O_3^{2-} \rightarrow S_4O_6^{2-} + 2e^-$$
$$I_2 + 2e^- \rightarrow 2I^-$$

Free iodine or substances which liberate iodine from acidified potassium iodide solution can be studied by titration with sodium thiosulphate [thiosulphate(VI)] in neutral solution. Iodine dissolves in potassium iodide solution giving the triiodide ion.

$$I_2 + I^- \rightleftharpoons I_3^-$$

The relative molecular mass of sodium thiosulphate-5-water is 248.2, so the mass required for 250 cm^3 of 0.1M solution $\approx$ 6.2 g: the equation involves two electrons and two thiosulphate ions and so the concentration for comparison is 0.1M. Sodium thiosulphate is not a primary standard so potassium permanganate [manganate(VII)], potassium dichromate [dichromate(VI)] or potassium iodate [iodate(V)] is used to standardize it. The disadvantages of sodium thiosulphate are given below.

1) Bacteria requiring sulphur may attack it.
2) Carbon dioxide in distilled water may convert thiosulphate into sulphate ions.
3) Atmospheric oxidation may convert thiosulphate into sulphate ions.
4) Any traces of copper in the distilled water will catalyze its decomposition.

Although iodine is not a primary standard because of its volatility it can be used to make a standard solution with very little error. Iodine solution can be used for back titration estimations.

The colour of a highly dilute iodine solution is too slight for any change to serve as an indication of the end point of the reaction. Soluble starch is used as the indicator: it gives an intense blue colour with iodine even at low concentrations. The blue coloured compound is a tunnel clathrate formed by the starch molecule coiling itself round the iodine molecules; the reaction is easily reversible.

6.29 The Standardization of Sodium Thiosulphate with Potassium Permanganate

1 $$2MnO_4^- + 16H^+ + 10I^- \rightarrow 2Mn^{2+} + 8H_2O + 5I_2$$
$$I_2 + 2S_2O_3^{2-} \rightarrow 2I^- + S_4O_6^{2-}$$

The permanganate solution must be standardized with an iron(II) salt or an oxalate (ethanedioate, see experiments 6.14 and 6.15).

Make up 250 cm^3 of 0.1M sodium thiosulphate solution.

2 Titrate 25 cm^3 portions of 0.02M potassium permanganate solution, to which about 20 cm^3 of 1M sulphuric acid and 10 cm^3 0.5M potassium iodide solution are added each time, with sodium thiosulphate solution. When the solution has changed from brown to pale yellow add about 1 cm^3 of starch solution as indicator and then continue adding sodium thiosulphate solution until the mixture suddenly becomes colourless. The starch should not be added at the beginning of the titration because in that case no indication of the end-point would be apparent.

3 Compare the concentration of the sodium thiosulphate solution obtained from the weighings with that obtained from the titrations.

6.30 The Standardization of Sodium Thiosulphate with Potassium Iodate [Iodate(V)]

1 $$IO_3^- + 6H^+ + 5I^- \rightarrow 3I_2 + 3H_2O$$
$$I_2 + 2S_2O_3^{2-} \rightarrow 2I^- + S_4O_6^{2-}$$

The relative molecular mass of potassium iodate is 214.0 so the mass to be taken for 250 cm^3 of 0.016M solution $\approx$ 0.9 g.

2 Titrate 25 cm^3 portions of potassium iodate solution to which about 20 cm^3 of 1M sulphuric acid and about 10 cm^3 0.5M potassium iodide solution are added each time, with sodium thiosulphate solution, adding starch solution as the indicator towards the end-point.

3 Calculate the concentration of the sodium thiosulphate solution.

6.31 The Determination of the Proportion of Copper in a Salt

1 $2Cu^{2+} + 4I^- \rightarrow 2CuI\downarrow + I_2$
   Off-white
   $I_2 + 2S_2O_3^{2-} \rightarrow 2I^- + S_4O_6^{2-}$
   The relative molecular mass of copper(II) sulphate-5-water is 249.7 so the mass required for $250\,cm^3$ of 0.1M solution $\approx 6.2\,g$.

2 Copper sulphate solution is acidic by hydrolysis and so each $25\,cm^3$ portion taken for titration must be carefully neutralized. This is done by adding sodium carbonate solution until there is a slight precipitate and then dilute acetic (ethanoic) acid until the precipitate just dissolves.

3 To the neutralized copper sulphate solution add about $20\,cm^3$ 0.5 M potassium iodide solution and then titrate it with standard (e.g. 0.1M) thiosulphate solution adding starch solution as the indicator towards the end-point. An off-white precipitate remains in each case.

4 Calculate the percentage of copper in the crystals supplied and compare the value obtained with that calculated from the relative molecular mass.

6.32 The Determination of the Available Chlorine in a Bleaching Agent

1 Qualitative analysis shows that 'Domestos', 'Brobat', 'Parazone', 'Chloros', 'Milton', etc. consist mainly of sodium hypochlorite [chlorate(I)] and they will cause bleaching of fabric, destruction of germs etc. If their solutions are acidified chlorine is released and they may be compared by the extent to which chlorine is available. In the presence of potassium iodide solution iodine is preferentially released. For the purposes of the titration the bleach should be diluted ten times or as found by trial and error.

2 $Cl_2 + 2I^- \rightarrow I_2 + 2Cl^-$
   $I_2 + 2S_2O_3^{2-} \rightarrow 2I^- + S_4O_6^{2-}$
   Titrate $25\,cm^3$ portions of the diluted bleach, to which about $20\,cm^3$ of 1M sulphuric acid and $20\,cm^3$ 0.5M potassium iodide solution are added each time, with standard (e.g. 0.1M) thiosulphate solution adding starch solution as the indicator towards the end-point.

3 Calculate the percentage (m/V) of available chlorine in the original bleach.

6.33 The Determination of the Purity of a Sulphite [Sulphate(IV)]

1 $SO_3^{2-} + I_2 + H_2O \rightarrow SO_4^{2-} + 2H^+ + 2I^-$
   In excess
   $H^+ + HCO_3^- \rightarrow H_2O + CO_2\uparrow$
   $I_2 + 2S_2O_3^{2-} \rightarrow 2I^- + S_4O_6^{2-}$
   The relative molecular mass of sodium sulphite-7-water is 252.2 so the mass of crystals to be taken for $250\,cm^3$ of 0.05M solution $\approx 3.2\,g$.

2 To $50\,cm^3$ 0.05M iodine solution add (with shaking) $25\,cm^3$ of the sulphite solution and then two spatula loads of sodium hydrogencarbonate.

3 Titrate the excess iodine present with standard (e.g. 0.1M) thiosulphate solution using starch solution as the indicator towards the end-point.

4 Calculate the concentration of the residual iodine solution and hence the number of millimoles remaining. This is another example of a back titration. Calculate the percentage purity of the crystals.

5 Test the crystals or their solution for impurities, e.g. sulphate, chloride, etc., and devise explanations for those radicals found to be present.

6.34 The Determination of the Hydrogen Peroxide Content of a Hair Bleach

1 The hair bleach has been diluted five times for the purpose of this titration.
   $H_2O_2 + 2I^- + 2H^+ \rightarrow I_2 + 2H_2O$
   $I_2 + 2S_2O_3^{2-} \rightarrow 2I^- + S_4O_6^{2-}$

2 To $25\,cm^3$ of the diluted bleach add about $50\,cm^3$ of 1M sulphuric acid, about $20\,cm^3$ 0.5M potassium iodide solution and about $0.5\,cm^3$ ammonium molybdate [molybdate(VI)] solution (5% m/V, i.e. $\approx 0.1M$). Wait five minutes for the reaction to be completed (the molybdate acts as a catalyst); meanwhile put out the solutions for the next titration.

3 Titrate the iodine liberated with standard (e.g. 0.1M) thiosulphate solution using starch solution as an indicator towards the end-point.

4 Calculate the concentration of the hair bleach and hence the percentage (m/V) of hydrogen peroxide in it.

5 Calculate the volume strength of the hydrogen peroxide using the equation for its thermal or catalytic decomposition. See experiment 6.18(4).

6.35 The Determination of the Solubility of Sulphur Dioxide in Water

1 The saturated solution of sulphur dioxide in water at room temperature should be diluted 50 times.

$$SO_2 + 2H_2O + I_2 \quad \rightarrow \quad SO_4^{2-} + 4H^+ + 2I^-$$
In excess

$$H^+ + HCO_3^- \quad \rightarrow \quad H_2O + CO_2 \uparrow$$
$$I_2 + 2S_2O_3^{2-} \quad \rightarrow \quad 2I^- + S_4O_6^{2-}$$

2 To $50\,cm^3$ $0.05\,M$ iodine solution add, with shaking, $25\,cm^3$ of the diluted sulphur dioxide solution and then two spatula loads of sodium hydrogencarbonate.

3 Titrate the excess iodine present with standard (e.g. 0.1M) thiosulphate solution using starch solution as the indicator towards the end-point.

4 Calculate the concentration of the residual iodine solution and hence the number of millimoles remaining. This is another example of a back titration. Calculate the solubility of the sulphur dioxide in terms of the mass dissolved in $1\,dm^3$ of solution.

6.36 The Determination of the Relative Atomic Mass of Chromium

1 $$Cr_2O_7^{2-} + 14H^+ + 6I^- \rightarrow 3I_2 + 2Cr^{3+} + 7H_2O$$
$$I_2 + 2S_2O_3^{2-} \rightarrow 2I^- + S_4O_6^{2-}$$
A solution of potassium dichromate containing $5\,g/dm^3$ of the crystals is supplied. It is assumed that the formula of potassium dichromate is $K_2Cr_2O_7$ and that the relative atomic masses of potassium and oxygen are 39.10 and 16.00 respectively.

2 To $25\,cm^3$ of the dichromate solution add about $75\,cm^3$ of 1M sulphuric acid, about $10\,cm^3$ 0.5M potassium iodide solution, about $10\,cm^3$ glacial acetic (ethanoic) acid and two drops of 1M copper(II) sulphate solution (a catalyst). Stir the contents of the flask vigorously and then add four spatula loads of sodium hydrogencarbonate.
$$H^+ + HCO_3^- \rightarrow H_2O + CO_2 \uparrow$$

3 Titrate the iodine liberated with standard (e.g. 0.1M) thiosulphate solution using starch solution as the indicator towards the end-point of the reaction.

4 Calculate the concentration of the dichromate solution and hence the relative atomic mass of chromium.

Redox Titrations: (d) Cerium

Introduction

In acid solution cerium(IV) salts are powerful oxidizing agents.

$$Ce^{4+} + e^- \rightarrow Ce^{3+}$$
Intense yellow   Pale yellow

N-phenylanthranilic (N-phenyl-2-aminobenzoic) acid is usually employed as the indicator: the solution becomes red-violet when there is an excess of the oxidizing agent. Two points in favour of using cerium(IV) salts are given below.

1) They are very stable when stored in solution.
2) They can be used in the presence of hydrochloric acid.

6.37 The Determination of the Purity of a Cerium(IV) Salt

1   $Ce^{4+} + Fe^{2+} \rightarrow Ce^{3+} + Fe^{3+}$

The relative molecular mass of cerium(IV) sulphate is 332.2. The salt supplied may be hydrated so it must be heated to constant mass. The mass required for 250 cm^3 of 0.1M solution (if pure) $\approx$ 8.4 g. In making up the solution about 50 cm^3 of 1M sulphuric acid should be included to reduce hydrolysis; warming the solution may facilitate dissolution but the solution must be carefully cooled to room temperature before it is put in the measuring flask.

2   Titrate 10 cm^3 (or 25 cm^3 if the titres are low) portions of standard iron(II) solution (e.g. 0.1M ammonium iron(II) sulphate solution) to which about 50 cm^3 of 1M sulphuric acid are added each time, using N-phenylanthranilic acid as the indicator.

3   Calculate the concentration of the cerium(IV) solution and hence the percentage purity of the anhydrous salt.

6.38 The Determination of the Purity of a Copper(I) Salt

1   $Fe^{3+} + Cu^+ \rightarrow Fe^{2+} + Cu^{2+}$
$Ce^{4+} + Fe^{2+} \rightarrow Ce^{3+} + Fe^{3+}$

Some copper(I) chloride can be prepared (see experiment 2.25) and its purity found. The copper (I) salt reduces an iron(III) salt to the iron(II) state and then the latter is titrated with the cerium(IV) solution. A suitable iron(III) solution can be prepared by dissolving $\approx$ 10 g of ammonium iron(III) sulphate-12-water in 50 cm^3 of 1M sulphuric acid.

2   Put about 20 cm^3 of the iron(III) salt solution in a conical flask and add about 0.2 g copper(I) chloride. Swirl the flask vigorously and add a few drops of N-phenylanthranilic acid as indicator. Titrate the mixture with a standard cerium(IV) (e.g. 0.03 M) solution.

3   Calculate the concentration of the iron(II) solution and hence the percentage purity of the copper(I) chloride.

Silver Nitrate Titrations

Introduction

$$Ag^+ + X^- \rightarrow AgX\downarrow$$
$$(X = Cl, Br \text{ or } I)$$

Silver nitrate (relative molecular mass 169.9) and sodium chloride (relative molecular mass 58.44) are obtainable in a high state of purity and so either can be employed as the standardizing agent. The distilled water used to make up the solutions must be of a high quality; 0.1M solutions are used.

The titrations should not be performed in direct sunlight so that photochemical changes may be avoided. Vigorous shaking of the flask is advantageous and coagulation (lump formation) of the precipitate usually occurs just before the end-point. There are various ways of finding the end-point of the precipitation.

1 The nephelometric method
In this method no indicator is added: the precipitate is allowed to settle and when another drop of solution from the burette causes no further precipitation then the end-point has been reached. This method is slow and tedious.

2 Mohr's method (1856)
Potassium chromate (about 1 cm³ of the 0.1M solution) is added as the indicator. Silver chromate [chromate(VI)] is more soluble than silver chloride so no red precipitate appears in addition to the white one until all the latter has formed. The disadvantage of this method is that it cannot be used directly in solutions that are not neutral: in acidic solutions silver chromate is not precipitated at all and in alkaline solutions silver oxide would be precipitated.

3 Volhard's method (1874)
This method can be employed in acid solution: it was originally devised to estimate silver dissolved in nitric acid. An excess of standard silver nitrate solution is added to the halide solution and the excess found by titration with a thiocyanate using an iron(III) salt as the indicator; a red coloration is given at the end-point.
$$Fe^{3+} + SCN^- \rightarrow [FeSCN]^{2+}$$
The disadvantage is that potassium and ammonium thiocyanates are not primary standards. Dilute nitric acid is added to reduce the hydrolysis of the iron(III) salt. It is advantageous but not essential to filter off the precipitate of silver halide obtained initially.

4 Fajan's method (1923)
The colour of an adsorption indicator depends on the nature of the surface of the precipitate upon which it is adsorbed. These indicators can be used in slightly acidic solutions and for mixtures of halides. Silver chloride usually forms as a colloid which coagulates towards the end-point. The crystal structure is face centred cubic with respect to each ion (like sodium chloride) and in two dimensions can be represented as (a). Before the end-point there is an excess of chloride ions and so the precipitate tends to adsorb these: the precipitate is therefore negatively charged (b). After the end-point the particles tend to adsorb silver cations and so become positively charged (c).

Thus the electrical field in the neighbourhood of the adsorbed indicators changes at the end-point and the indicator changes colour. Fluorescein which can be used for the three halides in neutral solution or in the presence of sodium acetate changes from a pale yellow-green to pink; eosin (tetrabromofluorescein), which can be used only for bromides and iodides, changes from pale pink to deep pink. Eosin can be used in acid solution providing the concentration of hydrogen ions is less than 0.1M.

A similar titration is that of lead nitrate with potassium hexacyanoferrate(II) in neutral solution using alizarin as the indicator. The colour change is from pale orange to pink.

On the grounds of expense titrations are done often using only 10 cm³ of solution rather than 25 cm³. The precipitates of silver chloride should be collected and sent to Johnson Matthey & Co. Ltd., Commercial Refining Department, Stockingswater Lane, Brimsdown, Enfield, Middlesex so that the silver can be regained when 1 kg has been accumulated.

6.39 The Reaction of Silver Nitrate with Sodium Chloride

1 $Ag^+ + Cl^- \rightarrow AgCl\downarrow$
Read the specifications for the silver nitrate and sodium chloride supplied to the laboratory. It may be found that the silver nitrate is of higher purity than the sodium chloride. Make up 250 cm³ of 0.1M solutions of each.

2 Titrate 10 cm³ portions of sodium chloride solution with the silver nitrate solution using potassium chromate [chromate(VI)] as the indicator.

3 Repeat (2) but using fluorescein as the indicator.

4 Compare the concentration of the silver nitrate solution calculated from the weighings with that obtained from the titration. Do the results agree with the manufacturer's specification?

183

6.40 The Standardization of Hydrochloric Acid with Silver Nitrate

1 $CaCO_3 + 2H^+ \rightarrow Ca^{2+} + H_2O + CO_2\uparrow$
   $Ag^+ + Cl^- \rightarrow AgCl\downarrow$

   A suitable sample of hydrochloric acid for studying is that supplied to the laboratory (probably 35% m/V) diluted 100 times.

2 To $10\,cm^3$ of the diluted acid add about 1 g of calcium carbonate powder and then titrate with silver nitrate using a suitable indicator.

3 Calculate the concentration of the hydrochloric acid supplied to the laboratory.

6.41 The Analysis of the Catholyte from a Diaphragm Cell

1 $Ag^+ + Cl^- \rightarrow AgCl\downarrow$
   $HCl + NaOH \rightarrow NaCl + H_2O$

   At the cathode in a diaphragm cell electrolysis of the sodium chloride solution yields sodium hydroxide solution but the conversion is not complete. The catholyte (solution near cathode after electrolysis) has been diluted 25 times for the titration.

2 To $10\,cm^3$ of the diluted solution add about $2\,cm^3$ of 1M nitric acid and then excess calcium carbonate (there must be some effervescence and then a white suspension showing an excess of calcium carbonate). Next add a suitable indicator and titrate with standard (e.g. 0.1M) silver nitrate solution.

3 Titrate a new $10\,cm^3$ portion of the catholyte with standard (e.g. 0.1M) hydrochloric acid using a suitable indicator.

4 Calculate the percentage (m/V) of the alkali and the salt in the original catholyte.

6.42 The Determination of the Relative Molecular Mass of Potassium Bromide

1 $Ag^+ + Br^- \rightarrow AgBr\downarrow$
   Titrate $25\,cm^3$ of potassium bromide solution $(10\,g/dm^3)$ with standard (e.g. 0.1M) silver nitrate solution using eosin as the indicator. Vigorous shaking of the flask is essential.

2 Calculate the concentration of the potassium bromide solution and hence the relative molecular mass of the solute.

6.43 The Determination of the Concentration of a Solution of a Mercury(I) Salt

1 $Hg_2^{2+} + 2Br^- \rightarrow Hg_2Br_2\downarrow$
   Using a burette, put $24\,cm^3$ mercury(I) nitrate solution into a conical flask, then add $25\,cm^3$ potassium bromide solution $(10\,g/dm^3)$ and shake the flask vigorously. Filter off the precipitate and wash it thoroughly with cold water, adding the rinsings to the original filtrate.

2 $Ag^+ + Br^- \rightarrow AgBr\downarrow$
   Titrate the filtrate and rinsings with standard (e.g. 0.1M) silver nitrate solution using eosin as the indicator.

3 Calculate the concentration (in g/dm^3) of crystalline mercury(I) nitrate, $Hg_2(NO_3)_2 \cdot 2H_2O$, in the solution.

6.44 The Standardization of Potassium Thiocyanate with Silver Nitrate

1 $Ag^+ + SCN^- \rightarrow AgSCN\downarrow$
    A chain molecule

The relative molecular mass of potassium thiocyanate is 97.18 and hence the mass to take to make up 250 cm^3 of 0.1M solution $\approx$ 2.4 g.

2 Titrate 10 cm^3 portions of standard (e.g. 0.1M) silver nitrate solution, to which about 10 cm^3 of 1M nitric acid and 1 cm^3 of an iron(III) solution (e.g. 0.5M ammonium iron(III) sulphate-12-water solution) are added each time, with the thiocyanate solution. At the end-point a red coloration is obtained, thorough shaking being essential.
    $Fe^{3+} + SCN^- \rightarrow [FeSCN]^{2+}$

3 Calculate the concentration of the thiocyanate solution and hence the percentage purity of the potassium thiocyanate.

6.45 The Standardization of Hydrochloric Acid with Potassium Thiocyanate

1 $Ag^+ + Cl^- \rightarrow AgCl\downarrow$
    Excess

    $Ag^+ + SCN^- \rightarrow AgSCN\downarrow$

A suitable sample of hydrochloric acid for studying is that supplied to the laboratory (probably 35% m/V) diluted 100 times.

2 Titrate 25 cm^3 of standard (e.g. 0.1M) silver nitrate solution to which 10 cm^3 of the diluted hydrochloric acid, about 10 cm^3 of 1M nitric acid and about 2 cm^3 nitrobenzene (this coats the precipitate and prevents it redissolving) are added each time, with potassium thiocyanate using an iron(III) salt as indicator.

3 Calculate the concentration and hence the percentage concentration (m/V) of the hydrochloric acid supplied to the laboratory.

Introduction

In these titrations it is usual for a metal ion to become surrounded by other species (ligands) which may be molecules or ions. As a general equation

$$M + nL \rightleftharpoons ML_n$$

may be written for which the equilibrium (or stability) constant is

$$K = \frac{[ML_n]}{[M][L]^n}$$

In the case of a multidentate ligand, i.e. one able to attach itself at several points to the central atom, the process is known as chelation.

Ethylenediaminetetraacetic acid (edta) or systematically bis[di(carboxymethyl)amino] ethane is such a substance having the formula

$$\begin{array}{l} CH_2 N(CH_2 COOH)_2 \\ | \\ CH_2 N(CH_2 COOH)_2 \end{array}$$

It was first used by Swarzenbach (1945). Some values of $\log_{10} K$, at 293 K (20°C), for edta with typical metal ions are as follows:

Mg^{2+}	8.7	Fe^{2+}	14.3	Cu^{2+}	18.8	Al^{3+}	16.3
Ca^{2+}	10.7	Fe^{3+}	25.1	Zn^{2+}	16.7	Pb^{2+}	18.0

Edta is usually employed as its disodium salt which has two molecules of water of crystallization: it has a relative molecular mass of 372.2 and is usually employed in 0.01M solution. The edta crystals must be carefully dried before use.

The sodium salt may be designated as $Na_2 H_2 E$ because it gives 1:1 complexes with many metal ions in solution forming strainless five-membered rings on chelation.

The indicator is a substance sensitive to metal ions which will also chelate giving a different coloured substance from the free indicator, e.g. Eriochrome (Solochrome) Black T which, to give it its full name, is sodium(I) 1-(1-hydroxy-2-naphthylazo)-6-nitro naphth-2-ol-4-sulphonate.

edta added are of the same pattern as the more familiar pH against volume of acid titration curves. The greater the stability constant of the chelate the sharper the endpoint at a given pH.

The indicator may be used in neutral or alkaline solution and is most effective if the pH is maintained at about 10: it may be represented as $H_2 T^-$, its phenolic groups being the most important. The indicator must react with the metal ion as soon as the edta-metal ion reaction is complete.

$$M^{2+} + H_2 T^- \rightarrow MT^- + 2H^+$$
$$\text{Blue} \qquad \quad \text{Red}$$

$$M^{2+} + H_2 E^{2-} \rightarrow EM^{2-} + 2H^+$$

As the titration proceeds the pH decreases and this may affect the stability of the chelate, e.g. for calcium and magnesium the pH must be kept above 7.5 but for other metals such as copper the pH may be as low as 3.5. In very alkaline solutions the solubility product of the metal hydroxide may be exceeded. Graphs of pM, i.e. $-\log_{10} [M^{n+}]$, plotted against the volume of

6.46 The Determination of the Total Hardness of Water

1 $M^{2+} + H_2E^{2-} \rightarrow EM^{2-} + 2H^+$
   edta

This method is employed if magnesium and calcium are responsible for the hardness of the water. A buffer solution is made up by adding 7 g ammonium chloride to 57 cm^3 concentrated ammonia solution and making up the total volume to 100 cm^3 with water.

---

2 Titrate 50 cm^3 portions of tap-water, to which about 1 cm^3 of buffer solution are added each time, with standard (e.g. 0.01M) edta solution using about 0.5 cm^3 Eriochrome Black T as indicator. The change in colour at the end-point is from red to blue.

---

3 Calculate the concentration of the metal ions in the tap-water and hence quote the hardness in mg/kg (parts per million) of calcium carbonate in water.

6.47 The Determination of the Proportion of Aluminium in a Salt

1 $Al^{3+} + H_2E^{2-} \rightarrow AlE^- + 2H^+$
   edta in
   excess

A suitable salt to study is aluminium potassium sulphate-12-water, the relative molecular mass of which is 474.4, so the mass required for 250 cm^3 of 0.01M solution $\approx$ 1.2 g.

A 0.01M zinc solution is also required for back titration of the excess edta: this may be made by making $\approx$ 0.7 g of zinc sulphate-7-water (relative molecular mass 287.5) into 250 cm^3 aqueous solution.
$Zn^{2+} + H_2E^{2-} \rightarrow EZn^{2-} + 2H^+$

---

2 To 25 cm^3 of the aluminium salt solution add 50 cm^3 of 0.01M edta solution and adjust the pH to between 7 and 8 by adding dilute ammonia solution, then boil the solution for a few minutes to ensure complete complexation of the aluminium ions.

---

3 Cool the solution to room temperature and again adjust the pH to between 7 and 8. Add 0.5 cm^3 of Eriochrome Black T as indicator and titrate the mixture rapidly with the standard zinc solution. The colour change at the end-point is from blue to red.

---

4 Calculate the concentration of the aluminium solution and hence the percentage of aluminium in the salt. Compare the value obtained by titration with that obtained by calculation using a table of relative atomic masses.

Gravimetric Analysis

Introduction

In gravimetric analysis a sample of the material is weighed out and then brought into solution by physical or chemical means. After the initial weighing, the quantities of materials — providing that they are in excess of that required for a given reaction — may not be crucial. Finally a precipitate is obtained and separated off. The precipitate is carefully dried: its composition may be known fairly accurately and it may be stable enough to be weighed directly, otherwise it is ignited and converted to a suitable compound. The relative atomic mass table is necessary for calculations.

6.48 The Determination of the Proportion of Aluminium in a Salt

1   A suitable compound to study is aluminium potassium sulphate-12-water: about 1.5 g of it should be weighed out accurately.

2   Dissolve the crystals in about $100 \, cm^3$ of distilled water and $5 \, cm^3$ of 1M sulphuric acid in a $250 \, cm^3$ beaker. The acid is added to prevent premature precipitation of aluminium hydroxide formed by hydrolysis.

3   Heat the solution to the boiling-point and then add dilute ammonia solution until no further precipitation occurs: about $25 \, cm^3$ may be required.
$$Al^{3+} + 3OH^- \rightarrow Al(OH)_3 \downarrow$$
Boil the solution for five minutes to assist coagulation and then after allowing the precipitate to settle decant off the clear supernatant liquid through a 110 mm filter-paper in a 75 mm funnel. The solution in the filter-paper should never be allowed to rise nearer than 10 mm to the upper edge.

4   Now wash the precipitate on to the filter-paper and rinse it thoroughly with hot distilled water until there are no more sulphate ions in the filtrate. Then dry the precipitate in an oven set at 380K ($100°C$ or just above).

5   Weigh a crucible and then transfer the filter-paper and precipitate into it. Ignite the paper to destroy it. Heat the crucible strongly to decompose the aluminium hydroxide.
$$2Al(OH)_3 \rightarrow Al_2O_3 + 3H_2O \uparrow$$

6   Allow the crucible and aluminium oxide to cool; after the temperature has decreased to a reasonable value place in a desiccator. Find the mass of the crucible, filter-paper ash and aluminium oxide; the filter-paper gives a known mass of ash.

7   Using a table of relative atomic masses calculate the mass of aluminium in the aluminium oxide and hence the percentage of aluminium in the original salt. Compare the value obtained with that calculated from the formula of the salt.

6.49 The Determination of the Proportion of Sulphate in a Salt

1 A suitable compound to study is aluminium potassium sulphate-12-water: about 0.3 g of it should be weighed out accurately.

2 Dissolve the crystals in about $100 \, cm^3$ of distilled water and $5 \, cm^3$ of 1M hydrochloric acid in a $250 \, cm^3$ beaker. The acid is added to prevent precipitation of aluminium hydroxide formed by hydrolysis.

3 Add about $20 \, cm^3$ of hot 0.1M barium chloride solution until precipitation is complete.
$$Ba^{2+} + SO_4^{2-} \rightarrow BaSO_4\downarrow$$
Boil the solution for five minutes to assist coagulation and then, after allowing the precipitate to settle, decant off the clear supernatant liquid through a 110 mm filter-paper in a 75 mm funnel, as in experiment 6.48.

4 Now wash the precipitate on to the filter-paper and rinse it thoroughly with hot distilled water until there are no more chloride ions in the filtrate. Then dry the precipitate in an oven set at 380 K ($100°C$ or just above).

5 Weigh a crucible and then transfer the filter-paper and precipitate into it. Destroy the filter-paper by adding a few drops of concentrated nitric acid and heating the crucible with its contents.

6 Allow the crucible and barium sulphate to cool; after the temperature has decreased to a reasonable value place in a desiccator. Find the mass of the crucible, filter-paper ash and barium sulphate; the filter-paper gives a known mass of ash.

7 Using a table of relative atomic masses calculate the mass of sulphate in the barium sulphate and hence the percentage of sulphate in the original salt. Compare the value obtained with that calculated from the formula of the salt.

6.50 The Determination of the Proportion of Phosphate in a Fertilizer

1 A suitable compound to study is potassium dihydrogenphosphate: about 0.4 g of it should be weighed out accurately.

2 Dissolve the crystals in about $100 \, cm^3$ of distilled water and $3 \, cm^3$ of concentrated hydrochloric acid in the presence of a few drops of methyl orange.

3 Make up a fresh solution of 'magnesia mixture' as follows: dissolve 1.25 g magnesium chloride and 2.5 g ammonium chloride in $10 \, cm^3$ of distilled water and add enough dilute ammonia solution to make the solution alkaline, filtering if necessary; acidify the solution by adding concentrated hydrochloric acid dropwise and make up the total volume to about $25 \, cm^3$.

4 To the phosphate solution add the 'magnesia mixture' and then sufficient concentrated ammonia solution to turn the indicator yellow. This must all be done at room temperature. Stir the mixture well for ten minutes and then allow the precipitate to settle.
$$Mg^{2+} + NH_4^+ + PO_4^{3-} + 6H_2O \rightarrow MgNH_4PO_4 \cdot 6H_2O\downarrow$$

5 Weigh a 110 mm filter-paper and then decant off the clear supernatant liquid through it in a Buchner funnel. Wash the precipitate on to the filter-paper and rinse it thoroughly with cold ammonia solution until there are no more chloride ions in the filtrate.

6 Next dry the precipitate by pouring over it three $10 \, cm^3$ portions of ethanol followed by three $5 \, cm^3$ portions of diethyl ether (ethoxyethane). Draw air through the precipitate and paper for ten minutes and allow these solvents to vaporize. Put the filter-paper and funnel in a desiccator for 20 minutes. **Care:** diethyl ether is flammable.

7 Weigh the filter-paper and precipitate. This method is of moderate accuracy; greater accuracy can be achieved by ignition to magnesium pyrophosphate but this is not essential at this standard.

8 Using a table of relative atomic masses calculate the mass of phosphate in the precipitate and hence the percentage of phosphate in the original salt. Compare the value obtained with that calculated from the formula of the salt.

Logarithms

	0	1	2	3	4	5	6	7	8	9	1	2	3	4	5	6	7	8	9
10	0000	0043	0086	0128	0170	0212	0253	0294	0334	0374	4	8	12	17	21	25	29	33	37
11	0414	0453	0492	0531	0569	0607	0645	0682	0719	0755	4	8	11	15	19	23	26	30	34
12	0792	0828	0864	0899	0934	0969	1004	1038	1072	1106	3	7	10	14	17	21	24	28	31
13	1139	1173	1206	1239	1271	1303	1335	1367	1399	1430	3	6	10	13	16	19	23	26	29
14	1461	1492	1523	1553	1584	1614	1644	1673	1703	1732	3	6	9	12	15	18	21	24	27
15	1761	1790	1818	1847	1875	1903	1931	1959	1987	2014	3	6	8	11	14	17	20	22	25
16	2041	2068	2095	2122	2148	2175	2201	2227	2253	2279	3	5	8	11	13	16	18	21	24
17	2304	2330	2355	2380	2405	2430	2455	2480	2504	2529	2	5	7	10	12	15	17	20	22
18	2553	2577	2601	2625	2648	2672	2695	2718	2742	2765	2	5	7	9	12	14	16	19	21
19	2788	2810	2833	2856	2878	2900	2923	2945	2967	2989	2	4	7	9	11	13	16	18	20
20	3010	3032	3054	3075	3096	3118	3139	3160	3181	3201	2	4	6	8	11	13	15	17	19
21	3222	3243	3263	3284	3304	3324	3345	3365	3385	3404	2	4	6	8	10	12	14	16	18
22	3424	3444	3464	3483	3502	3522	3541	3560	3579	3598	2	4	6	8	10	12	14	15	17
23	3617	3636	3655	3674	3692	3711	3729	3747	3766	3784	2	4	6	7	9	11	13	15	17
24	3802	3820	3838	3856	3874	3892	3909	3927	3945	3962	2	4	5	7	9	11	12	14	16
25	3979	3997	4014	4031	4048	4065	4082	4099	4116	4133	2	3	5	7	9	10	12	14	15
26	4150	4166	4183	4200	4216	4232	4249	4265	4281	4298	2	3	5	7	8	10	11	13	15
27	4314	4330	4346	4362	4378	4393	4409	4425	4440	4456	2	3	5	6	8	9	11	13	14
28	4472	4487	4502	4518	4533	4548	4564	4579	4594	4609	2	3	5	6	8	9	11	12	14
29	4624	4639	4654	4669	4683	4698	4713	4728	4742	4757	1	3	4	6	7	9	10	12	13
30	4771	4786	4800	4814	4829	4843	4857	4871	4886	4900	1	3	4	6	7	9	10	11	13
31	4914	4928	4942	4955	4969	4983	4997	5011	5024	5038	1	3	4	5	7	8	10	11	12
32	5051	5065	5079	5092	5105	5119	5132	5145	5159	5172	1	3	4	5	7	8	9	11	12
33	5185	5198	5211	5224	5237	5250	5263	5276	5289	5302	1	3	4	5	6	8	9	10	12
34	5315	5328	5340	5353	5366	5378	5391	5403	5416	5428	1	3	4	5	6	8	9	10	11
35	5441	5453	5465	5478	5490	5502	5514	5527	5539	5551	1	2	4	5	6	7	9	10	11
36	5563	5575	5587	5599	5611	5623	5635	5647	5658	5670	1	2	4	5	6	7	8	10	11
37	5682	5694	5705	5717	5729	5740	5752	5763	5775	5786	1	2	3	5	6	7	8	9	10
38	5798	5809	5821	5832	5843	5855	5866	5877	5888	5899	1	2	3	5	6	7	8	9	10
39	5911	5922	5933	5944	5955	5966	5977	5988	5999	6010	1	2	3	4	5	7	8	9	10
40	6021	6031	6042	6053	6064	6075	6085	6096	6107	6117	1	2	3	4	5	6	7	9	10
41	6128	6138	6149	6160	6170	6180	6191	6201	6212	6222	1	2	3	4	5	6	7	8	9
42	6232	6243	6253	6263	6274	6284	6294	6304	6314	6325	1	2	3	4	5	6	7	8	9
43	6335	6345	6355	6365	6375	6385	6395	6405	6415	6425	1	2	3	4	5	6	7	8	9
44	6435	6444	6454	6464	6474	6484	6493	6503	6513	6522	1	2	3	4	5	6	7	8	9
45	6532	6542	6551	6561	6571	6580	6590	6599	6609	6618	1	2	3	4	5	6	7	8	9
46	6628	6637	6646	6656	6665	6675	6684	6693	6702	6712	1	2	3	4	5	6	7	7	8
47	6721	6730	6739	6749	6758	6767	6776	6785	6794	6803	i	2	3	4	5	5	6	7	8
48	6812	6821	6830	6839	6848	6857	6866	6875	6884	6893	1	2	3	4	4	5	6	7	8
49	6902	6911	6920	6928	6937	6946	6955	6964	6972	6981	1	2	3	4	4	5	6	7	8
50	6990	6998	7007	7016	7024	7033	7042	7050	7059	7067	1	2	3	3	4	5	6	7	8
51	7076	7084	7093	7101	7110	7118	7126	7135	7143	7152	1	2	3	3	4	5	6	7	8
52	7160	7168	7177	7185	7193	7202	7210	7218	7226	7235	1	2	2	3	4	5	6	7	7
53	7243	7251	7259	7267	7275	7284	7292	7300	7308	7316	1	2	2	3	4	5	6	6	7
54	7324	7332	7340	7348	7356	7364	7372	7380	7388	7396	1	2	2	3	4	5	6	6	7

	0	1	2	3	4	5	6	7	8	9	1	2	3	4	5	6	7	8	9
55	7404	7412	7419	7427	7435	7443	7451	7459	7466	7474	1	2	2	3	4	5	5	6	7
56	7482	7490	7497	7505	7513	7520	7528	7536	7543	7551	1	2	2	3	4	5	5	6	7
57	7559	7566	7574	7582	7589	7597	7604	7612	7619	7627	1	2	2	3	4	5	5	6	7
58	7634	7642	7649	7657	7664	7672	7679	7686	7694	7701	1	1	2	3	4	4	5	6	7
59	7709	7716	7723	7731	7738	7745	7752	7760	7767	7774	1	1	2	3	4	4	5	6	7
60	7782	7789	7796	7803	7810	7818	7825	7832	7839	7846	1	1	2	3	4	4	5	6	6
61	7853	7860	7868	7875	7882	7889	7896	7903	7910	7917	1	1	2	3	4	4	5	6	6
62	7924	7931	7938	7945	7952	7959	7966	7973	7980	7987	1	1	2	3	3	4	5	6	6
63	7993	8000	8007	8014	8021	8028	8035	8041	8048	8055	1	1	2	3	3	4	5	6	6
64	8062	8069	8075	8082	8089	8096	8102	8109	8116	8122	1	1	2	3	3	4	5	5	6
65	8129	8136	8142	8149	8156	8162	8169	8176	8182	8189	1	1	2	3	3	4	5	5	6
66	8195	8202	8209	8215	8222	8228	8235	8241	8248	8254	1	1	2	3	3	4	5	5	6
67	8261	8267	8274	8280	8287	8293	8299	8306	8312	8319	1	1	2	3	3	4	4	5	6
68	8325	8331	8338	8344	8351	8357	8363	8370	8376	8382	1	1	2	3	3	4	4	5	6
69	8388	8395	8401	8407	8414	8420	8426	8432	8439	8445	1	1	2	3	3	4	4	5	6
70	8451	8457	8463	8470	8476	8482	8488	8494	8500	8506	1	1	2	2	3	4	4	5	6
71	8513	8519	8525	8531	8537	8543	8549	8555	8561	8567	1	1	2	2	3	4	4	5	5
72	8573	8579	8585	8591	8597	8603	8609	8615	8621	8627	1	1	2	2	3	4	4	5	5
73	8633	8639	8645	8651	8657	8663	8669	8675	8681	8686	1	1	2	2	3	4	4	5	5
74	8692	8698	8704	8710	8716	8722	8727	8733	8739	8745	1	1	2	2	3	4	4	5	5
75	8751	8756	8762	8768	8774	8779	8785	8791	8797	8802	1	1	2	2	3	3	4	5	5
76	8808	8814	8820	8825	8831	8837	8842	8848	8854	8859	1	1	2	2	3	3	4	5	5
77	8865	8871	8876	8882	8887	8893	8899	8904	8910	8915	1	1	2	2	3	3	4	4	5
78	8921	8927	8932	8938	8943	8949	8954	8960	8965	8971	1	1	2	2	3	3	4	4	5
79	8976	8982	8987	8993	8998	9004	9009	9015	9020	9025	1	1	2	2	3	3	4	4	5
80	9031	9036	9042	9047	9053	9058	9063	9069	9074	9079	1	1	2	2	3	3	4	4	5
81	9085	9090	9096	9101	9106	9112	9117	9122	9128	9133	1	1	2	2	3	3	4	4	5
82	9138	9143	9149	9154	9159	9165	9170	9175	9180	9186	1	1	2	2	3	3	4	4	5
83	9191	9196	9201	9206	9212	9217	9222	9227	9232	9238	1	1	2	2	3	3	4	4	5
84	9243	9248	9253	9258	9263	9269	9274	9279	9284	9289	1	1	2	2	3	3	4	4	5
85	9294	9299	9304	9309	9315	9320	9325	9330	9335	9340	1	1	2	2	3	3	4	4	5
86	9345	9350	9355	9360	9365	9370	9375	9380	9385	9390	1	1	2	2	3	3	4	4	5
87	9395	9400	9405	9410	9415	9420	9425	9430	9435	9440	0	1	1	2	2	3	3	4	4
88	9445	9450	9455	9460	9465	9469	9474	9479	9484	9489	0	1	1	2	2	3	3	4	4
89	9494	9499	9504	9509	9513	9518	9523	9528	9533	9538	0	1	1	2	2	3	3	4	4
90	9542	9547	9552	9557	9562	9566	9571	9576	9581	9586	0	1	1	2	2	3	3	4	4
91	9590	9595	9600	9605	9609	9614	9619	9624	9628	9633	0	1	1	2	2	3	3	4	4
92	9638	9643	9647	9652	9657	9661	9666	9671	9675	9680	0	1	1	2	2	3	3	4	4
93	9685	9689	9694	9699	9703	9708	9713	9717	9722	9727	0	1	1	2	2	3	3	4	4
94	9731	9736	9741	9745	9750	9754	9759	9764	9768	9773	0	1	1	2	2	3	3	4	4
95	9777	9782	9786	9791	9795	9800	9805	9809	9814	9818	0	1	1	2	2	3	3	4	4
96	9823	9827	9832	9836	9841	9845	9850	9854	9859	9863	0	1	1	2	2	3	3	4	4
97	9868	9872	9877	9881	9886	9890	9894	9899	9903	9908	0	1	1	2	2	3	3	4	4
98	9912	9917	9921	9926	9930	9934	9939	9943	9948	9952	0	1	1	2	2	3	3	4	4
99	9956	9961	9965	9969	9974	9978	9983	9987	9991	9996	0	1	1	2	2	3	3	4	4

Table of Relative Atomic Masses

Based on the assigned relative atomic mass of $^{12}C = 12$, the following values apply to elements as they exist in materials of terrestrial origin and to certain artificial elements.

Atomic Number	Name	Symbol	A_r	Atomic Number	Name	Symbol	A_r
89	Actinium	Ac	(227)	80	Mercury	Hg	200.6
13	Aluminium	Al	26.98	42	Molybdenum	Mo	95.94
95	Americium	Am	(243)	60	Neodymium	Nd	144.2
51	Antimony	Sb	121.8	10	Neon	Ne	20.18
18	Argon	Ar	39.95	93	Neptunium	Np	(237)
33	Arsenic	As	74.92	28	Nickel	Ni	58.70
85	Astatine	At	(210)	41	Niobium	Nb	92.91
56	Barium	Ba	137.3	7	Nitrogen	N	14.01
97	Berkelium	Bk	(247)	102	Nobelium	No	(259)
4	Beryllium	Be	9.012	76	Osmium	Os	190.2
83	Bismuth	Bi	209.0	8	Oxygen	O	16.00
5	Boron	B	10.81	46	Palladium	Pd	106.4
35	Bromine	Br	79.90	15	Phosphorus	P	30.97
48	Cadmium	Cd	112.4	78	Platinum	Pt	195.1
55	Caesium	Cs	132.9	94	Plutonium	Pu	(244)
20	Calcium	Ca	40.08	84	Polonium	Po	(209)
98	Californium	Cf	(251)	19	Potassium	K	39.10
6	Carbon	C	12.01	59	Praseodymium	Pr	140.9
58	Cerium	Ce	140.1	61	Promethium	Pm	(145)
17	Chlorine	Cl	35.45	91	Protactinium	Pa	231.0
24	Chromium	Cr	52.00	88	Radium	Ra	226.0
27	Cobalt	Co	58.93	86	Radon	Rn	(222)
29	Copper	Cu	63.55	75	Rhenium	Re	186.2
96	Curium	Cm	(247)	45	Rhodium	Rh	102.9
66	Dysprosium	Dy	162.5	37	Rubidium	Rb	85.47
99	Einsteinium	Es	(254)	44	Ruthenium	Ru	101.1
68	Erbium	Er	167.3	62	Samarium	Sm	150.4
63	Europium	Eu	152.0	21	Scandium	Sc	44.96
100	Fermium	Fm	(257)	34	Selenium	Se	78.96
9	Fluorine	F	19.00	14	Silicon	Si	28.09
87	Francium	Fr	(223)	47	Silver	Ag	107.9
64	Gadolinium	Gd	157.3	11	Sodium	Na	22.99
31	Gallium	Ga	69.72	38	Strontium	Sr	87.62
32	Germanium	Ge	72.59	16	Sulphur	S	32.06
79	Gold	Au	197.0	73	Tantalum	Ta	180.9
72	Hafnium	Hf	178.5	43	Technetium	Tc	(97)
2	Helium	He	4.003	52	Tellurium	Te	127.6
67	Holmium	Ho	164.9	65	Terbium	Tb	158.9
1	Hydrogen	H	1.008	81	Thallium	Tl	204.4
49	Indium	In	114.8	90	Thorium	Th	232.0
53	Iodine	I	126.9	69	Thulium	Tm	168.9
77	Iridium	Ir	192.2	50	Tin	Sn	118.7
26	Iron	Fe	55.85	22	Titanium	Ti	47.90
36	Krypton	Kr	83.80	74	Tungsten	W	183.8
57	Lanthanum	La	138.9	104	Unnilquadium	Unq	–
103	Lawrencium	Lr	(260)	92	Uranium	U	238.0
82	Lead	Pb	207.2	23	Vanadium	V	50.94
3	Lithium	Li	6.941	54	Xenon	Xe	131.3
71	Lutetium	Lu	175.0	70	Ytterbium	Yb	173.0
12	Magnesium	Mg	24.30	39	Yttrium	Y	88.91
25	Manganese	Mn	54.94	30	Zinc	Zn	65.38
101	Mendelevium	Md	(258)	40	Zirconium	Zr	91.22

Reproduced with permission from the International Union of Pure and Applied Chemistry.

Values are given to four significant figures.

Values in brackets refer to the longest-lived nuclides.